SpringerBriefs in Computer Science

SpringerBriefs present concise summaries of cutting-edge research and practical applications across a wide spectrum of fields. Featuring compact volumes of 50 to 125 pages, the series covers a range of content from professional to academic.

Typical topics might include:

- A timely report of state-of-the art analytical techniques
- A bridge between new research results, as published in journal articles, and a contextual literature review
- A snapshot of a hot or emerging topic
- An in-depth case study or clinical example
- A presentation of core concepts that students must understand in order to make independent contributions

Briefs allow authors to present their ideas and readers to absorb them with minimal time investment. Briefs will be published as part of Springer's eBook collection, with millions of users worldwide. In addition, Briefs will be available for individual print and electronic purchase. Briefs are characterized by fast, global electronic dissemination, standard publishing contracts, easy-to-use manuscript preparation and formatting guidelines, and expedited production schedules. We aim for publication 8–12 weeks after acceptance. Both solicited and unsolicited manuscripts are considered for publication in this series.

**Indexing: This series is indexed in Scopus, Ei-Compendex, and zbMATH **

Zhi Yan

Robot Perception and Learning

A Human-Aware Navigation and Long-Term Autonomy Perspective

Zhi Yan
Computer Science and Systems
Engineering Laboratory (U2IS)
ENSTA—Institut Polytechnique de Paris
Palaiseau, France

ISSN 2191-5768　　　　　　　ISSN 2191-5776　(electronic)
SpringerBriefs in Computer Science
ISBN 978-981-96-7093-2　　　ISBN 978-981-96-7094-9　(eBook)
https://doi.org/10.1007/978-981-96-7094-9

© The Editor(s) (if applicable) and The Author(s), under exclusive license to Springer Nature Singapore Pte Ltd. 2026

This work is subject to copyright. All rights are solely and exclusively licensed by the Publisher, whether the whole or part of the material is concerned, specifically the rights of translation, reprinting, reuse of illustrations, recitation, broadcasting, reproduction on microfilms or in any other physical way, and transmission or information storage and retrieval, electronic adaptation, computer software, or by similar or dissimilar methodology now known or hereafter developed.
The use of general descriptive names, registered names, trademarks, service marks, etc. in this publication does not imply, even in the absence of a specific statement, that such names are exempt from the relevant protective laws and regulations and therefore free for general use.
The publisher, the authors and the editors are safe to assume that the advice and information in this book are believed to be true and accurate at the date of publication. Neither the publisher nor the authors or the editors give a warranty, expressed or implied, with respect to the material contained herein or for any errors or omissions that may have been made. The publisher remains neutral with regard to jurisdictional claims in published maps and institutional affiliations.

This Springer imprint is published by the registered company Springer Nature Singapore Pte Ltd.
The registered company address is: 152 Beach Road, #21-01/04 Gateway East, Singapore 189721, Singapore

If disposing of this product, please recycle the paper.

To Cindy, Tony and Fanfan

Foreword

I am delighted to introduce *Robot Perception and Learning: A Human-Aware Navigation and Long-Term Autonomy Perspective* by Dr. Zhi Yan. I have had the pleasure of knowing and working with Zhi for many years, and this book reflects his important contributions to robotics, both in theory and practice.

This work takes readers through Zhi's decade-long journey in robotics, from the early challenges of labeling data for cleaning robots to developing a strong framework for how robots can learn and adapt in complex environments with people. It provides a detailed overview of research aimed at improving robot perception and learning, focusing on human-aware navigation and long-term autonomy. Key topics include benchmarking methods, 3D lidar-based object detection and tracking, and the emerging field of robot online learning (ROL). The book highlights the importance of standardized benchmarks and open datasets, introduces new techniques for processing point cloud data, and explores ROL's role in helping robots learn over time. It also presents solutions for challenges like generating training samples automatically and mitigating catastrophic forgetting. The book applies these ideas to socially aware robot navigation. While challenges remain—such as missing ground truth data and engineering limitations—it offers insightful ideas for future research, like combining online learning with reinforcement learning, improving prediction skills, and addressing privacy concerns.

Overall, this book is a valuable resource for advancing robot perception and learning, especially in the context of human-aware, long-term autonomous systems. It is particularly useful for graduate students, researchers, and professionals looking to understand the key ideas shaping this fast-growing field.

Zhi's work, through his research and open-source contributions (available via the provided links), has continuously driven innovation in robotics. The same applies to this book, which I believe will be a great resource for anyone working on robot perception and learning. I highly recommend it to anyone interested in the future of intelligent autonomous robots.

Padua, Italy Nicola Bellotto
February 2025

Preface

In 2015, I received a task from Dr. Nicola Bellotto to let a professional cleaning robot identify the points representing humans in the point cloud generated by 3D lidar. Then, based on the research progress at that time, I developed a point cloud annotation tool, collected point cloud data, manually annotated thousands of point cloud frames, and then trained an SVM model, tested with the robot, tuned the training parameters, and trained the model again. Then suddenly the robot changed its deployment environment, and I needed to collect data again and annotate them again, over and over again. Later, in Cyprus, I complained to Nicola, why can't the robot learn these boring point clouds by itself? After returning to Lincoln, Dr. Tom Duckett joined the discussion, and finally in July 2016, we completed the draft of our first "robot online learning" paper together.

Since 2015, robot perception and learning have been the main line of my research. Ten years later, the results of my own research and those I supervised were recognized at the *Institut Polytechnique de Paris*, and I was finally enabled to obtain the *Habilitation à Diriger des Recherches* in France. To this end, I needed to prepare a thesis to systematically summarize my research since my doctoral defense. This led to the writing of this book.

The pursuit of truly intelligent robots capable of operating effectively in complex and unstructured environments has driven significant advancements in both perception and learning. While traditional robotics often relied on pre-programmed instructions and carefully engineered environments, the increasing demand for mobile robots that can adapt, interact, and learn from their environment and experience has propelled the field towards embodied intelligence. This paradigm emphasizes the crucial role of physical embodiment in shaping perception, action, and cognition, leading to more robust and adaptable robotic systems.

This book aims to provide an overview of the latest research at the intersection of these crucial areas, with a particular focus on 3D lidar-based perception and robot online learning methods, with downstream tasks including human-aware robot navigation and long-term robot autonomy. It is structured as follows: Chap. 1 introduces the background of the research, the content positioning, and some related open-source resources. Chapter 2 discusses some benchmarking issues related to the field of embodied intelligence and mobile robotics. Chapter 3 introduces robot perception, especially the object detection and tracking based on 3D lidar with contemporary characteristics. Chapter 4 introduces robot learning, especially robot online learning methods with strong embodied intelligence features. Chapter 5 summarizes the book and provides prospects for future research and application directions.

This book is intended for researchers and practitioners already familiar with the fundamentals of mobile robotics, including graduate students, Ph.D. candidates, post-doctoral researchers, and industry professionals working in related fields. This book is not a textbook in the traditional sense; it does not delve into basic introductory material. Instead, it focuses on presenting the core methodologies and principles that drive cutting-edge research, providing a structured overview of key concepts and recent developments. Detailed experimental results and specific implementation details are intentionally omitted to maintain a focus on the broader theoretical and methodological underpinnings, encouraging readers to consult the original research papers for in-depth analysis and practical implementation strategies.

This book would not have been possible without the contributions of numerous researchers whose work forms the foundation of this field. We gratefully acknowledge their invaluable contributions and hope that this book serves as a useful resource for those seeking to understand and contribute to the future of robot perception and learning. We also extend our sincere thanks to any reviewers who provided feedback on drafts of this material. We sincerely hope that this book inspires new research directions, facilitates collaboration within the community, and ultimately contributes to the advancement of embodied intelligence and mobile robotics.

Palaiseau, France   Zhi Yan
January 2025

Acknowledgements This book represents the culmination of years of research and collaboration, and I am deeply indebted to the many individuals and institutions who have contributed to its creation.

First and foremost, I express my sincere gratitude to my academic advisors, Dr. Nicola Bellotto and Dr. Tom Duckett, for their invaluable guidance, unwavering support, and insightful feedback on the research covered in this book. Their expertise and mentorship have been instrumental in shaping my understanding of robot perception and learning.

I am also deeply grateful to my Ph.D. students and postdoc at the University of Technology of Belfort-Montbéliard (UTBM) in France. In particular, I would like to thank Dr. Rui Yang for his contributions in mitigating the catastrophic forgetting problem in robot online learning and introducing the latter into autonomous driving, Dr. Iaroslav Okunevich for his contributions in combining deep neural networks with robot online learning and bringing them into socially-compliant navigation, and Dr. Tao Yang, with whom I collaborated on much of the work in this book.

I would like to thank my employer, ENSTA—Institut Polytechnique de Paris, without which this book would not have been possible.

This research was made possible by the generous funding provided by European Commission (H2020 FLOBOT, PI: Dr. Nicola Bellotto), UTBM and PHC Barrande Programme (for development of the EU long-term dataset), China Scholarship Council (for Dr. Rui Yang's Ph.D.), the Burgundy-Franche-Comté region (for Dr. Iaroslav Okunevich's Ph.D.), Renault (for Dr. Tao Yang's postdoc), and the French National Research Agency (for the NavWare project). I gratefully acknowledge their support, which enabled us to pursue this line of inquiry.

I would like to express my sincere appreciation to the team at Springer Nature for their professionalism, dedication, and commitment to bringing this book to a wider audience. Especially, I would like to thank Dr. Nick Zhu for his keen eye and thoughtful suggestions.

Finally, I wish to express my heartfelt gratitude to my family and friends for their unwavering support, patience, and understanding during the demanding process of writing this book. Their encouragement has been a constant source of motivation.

Competing Interests Zhi Yan has received research grants from the French National Research Agency (ANR), the PHC Barrande Programme, the Burgundy-Franche-Comté region, UTBM, Toyota, Renault, and NVIDIA.

The author has no conflicts of interest to declare that are relevant to the content of this book.

Contents

1	Introduction	1
	1.1 Research Background	1
	1.2 Research Positioning	2
	1.3 Related Open Source Projects	5
	1.4 Book Organization	7
	References	8
2	**Benchmarking in Mobile Robotics**	11
	2.1 Introduction	11
	2.2 Benchmarking Process	12
	2.2.1 Parameters	12
	2.2.2 Metrics	15
	2.2.3 Experimental Design	22
	2.3 Testbed Construction	24
	2.4 Dataset Building	25
	2.4.1 Hardware Platform	26
	2.4.2 Software Architecture	29
	2.4.3 Sensor Calibration	29
	2.4.4 Comparison Between Different Datasets	30
	2.5 Discussions	33
	2.5.1 Ranking-Driven Overfitting	33
	2.5.2 Benchmarking AI with AI	34
	2.5.3 Benchmarking Ethics	34
	2.5.4 Data Privacy	35
	References	36
3	**Robot Perception**	39
	3.1 Introduction	39
	3.2 3D Lidar	40
	3.2.1 Ranging Principle	41
	3.2.2 Scanning Architectures	42
	3.2.3 Physical Properties	43

		3.2.4	Data Representation	44
		3.2.5	Industrial Applications	45
	3.3	Object Detection in Point Clouds		47
		3.3.1	Point Cloud Segmentation	48
		3.3.2	Object Classification	54
	3.4	Multi-target Tracking in Point Clouds		58
		3.4.1	Data Association	59
		3.4.2	State Estimation	60
	3.5	Conclusion		61
	References			62
4	**Robot Learning**			67
	4.1	Introduction		67
	4.2	Why Study Robot Online Learning		68
	4.3	Challenges of Robot Online Learning		68
	4.4	Autonomous Sample Generation		72
		4.4.1	P–N Learning	73
		4.4.2	Knowledge Transfer	76
	4.5	Mitigation of Catastrophic Forgetting		80
		4.5.1	Learning Sample Extraction	80
		4.5.2	Short-Term Learning	83
		4.5.3	Long-Term Control	83
	4.6	Robot Online Learning for Navigation		86
		4.6.1	Basic Navigation Module	87
		4.6.2	Online Adaptation Module	88
	4.7	Conclusion		91
	References			92
5	**Conclusions and Perspectives**			95
	5.1	General Conclusions		95
	5.2	Research Perspectives		96
	References			97
Index				99

Acronyms

AI	Artificial Intelligence
AP	Average Precision
BEV	Bird's Eye View
BiT	Batch-incremental Training
CAN	Controller Area Network
CCD	Charge-Coupled Device
CMOS	Complementary Metal-Oxide-Semiconductor
DEW	Dynamic Expert Weights
DNN	Deep Neural Network
EKF	Extended Kalman Filter
EWMA	Exponentially Weighted Moving Average
FMCW	Frequency Modulated Continuous Wave
FN	False Negative
FoV	Field of View
FP	False Positive
GNN	Global Nearest Neighbor
GRU	Gated Recurrent Unit
HRI	Human-Robot Interaction
HraE	Hand-raised as Expert
IMU	Inertial Measurement Unit
IoU	Intersection over Union
JPDA	Joint Probabilistic Data Association
KPI	Key Performance Indicator
LSTM	Long Short-Term Memory
LSTOL	Long Short-Term Online Learning
MEMS	Micro-Electro-Mechanical Systems
MLP	Multi-Layer Perceptron
ODD	Operational Design Domain
OL	Online Learning
OPA	Optical Phased Array
ORF	Online Random Forest

PCD	Point Cloud Data
PCL	Point Cloud Library
RF	Random Forest
RL	Reinforcement Learning
ROL	Robot Online Learning
ROS	Robot Operating System
RTK	Real-Time Kinematic
SARL	Socially Attentive Reinforcement Learning
SLAM	Simultaneous Localization and Mapping
STEM	Science, Technology, Engineering, and Mathematics
SVM	Support Vector Machine
TN	True Negative
ToF	Time-of-Flight
TP	True Positive
UKF	Unscented Kalman Filter

Chapter 1
Introduction

Abstract This chapter establishes the research context of this book: embodied intelligence. It clarifies the research focus: leveraging computer science principles to develop mobile robots as productive tools for human society. Finally, it presents relevant open-source projects related to the content discussed herein.

Keywords Artificial intelligence · Embodied intelligence · Mobile robotics

1.1 Research Background

The continuous development of robot hardware and software is pushing the capabilities of robots to a new boundary, and this boundary is getting closer and closer to our imagination of robots, as portrayed in countless science fiction works. These robots serve our human society by interacting with us as well as the same environment we live in, and one of the enabling capabilities is the so-called *embodied intelligence*. So, what is embodied intelligence? Embodied intelligence refers to an intelligent system that perceives and acts based on the physical body, which obtains information, understands problems, makes decisions and implements actions through the interaction between the intelligent agent and the environment, thereby producing intelligent behavior and adaptability. The germination of its ideas can be traced back to the birth of Artificial Intelligence (AI), as Alan Turing wrote [1]:

> We may hope that machines will eventually compete with men in all purely intellectual fields. But which are the best ones to start with? Even this is a difficult decision. Many people think that a very abstract activity, like the playing of chess, would be best. It can also be maintained that it is best to provide the machine with the best sense organs that money can buy, and then teach it to understand and speak English. This process could follow the normal teaching of a child. Things would be pointed out and named, etc.

Playing chess can be seen as disembodied intelligence, such as what Deep Blue in the last century and AlphaGo in this century presented to people. The example of understanding and speaking English can be seen as embodied intelligence, corresponding to the topic that this book wants to discuss. These two different examples actually reveal the difference between agents and robots as per [2]:

© The Author(s), under exclusive license to Springer Nature Singapore Pte Ltd. 2026
Z. Yan, *Robot Perception and Learning*, SpringerBriefs in Computer Science,
https://doi.org/10.1007/978-981-96-7094-9_1

Fig. 1.1 Disembodied intelligence (left, *image source*: internet) versus embodied intelligence (right)

> We should be careful not to confuse multi-robot systems (MRS) with multi-agent systems (MAS) and distributed artificial intelligence (DAI), as MAS usually refers to traditional distributed computer systems in which individual nodes are stationary, while DAI is primarily concerned with problems involving software agents. In contrast, MRS involves mobile robots that can move in the physical world and must physically interact with each other.

Figure 1.1 gives an intuitive impression of the differences between disembodied intelligence[1] and embodied intelligence. In the meantime, Turing's second example also brings out two aspects that this book is concerned about: *(robot) perception and learning*.

So why do we, or rather robots, need embodied intelligence? First, as mentioned earlier, a robot is an entity that essentially needs to physically interact with the real world (assuming the world we are in is not virtual). From a philosophical point of view, individuals who lack embodied cognition lack actual existence, which is similar to Descartes's point of view "I think, therefore I am", which is opposite to Heidegger's later "I am, therefore I think". Second, in the physical world, context plays a crucial role in semantic understanding, and the lack of embodied intelligence will make it difficult for robots to make informed decisions based on situational awareness. Last but not least, the physical environment in which robots operate changes, slow like the seasons and fast like a pedestrian. The lack of embodied intelligence makes it difficult for robots to cope with changes in the real world.

1.2 Research Positioning

The research positioning of this book remains consistent at the macro level, that is, rooted in computer science to empower mobile robots with embodied intelligence and make them a productive tool for human society. But at the micro level it includes

---

[1] Image source: Google DeepMind (https://www.youtube.com/@Google_DeepMind).

1.2 Research Positioning

symbolic methods, statistical methods, connectionist methods, and activist methods. Specifically, this book investigates two issues. One is how to make a robot aware of the surrounding objects such as humans through embodied perception, including their detection and tracking [3], to provide necessary information for downstream tasks such as safe and even social robot navigation. The other is how to make a robot autonomously and spontaneously learn from its sensor data to understand its surroundings [4], such as humans in various forms, to achieve out-of-the-box usage and long-term robot autonomy.

Regarding the first issue, this book focuses on leveraging embodied non-visual sensors, especially 3D lidar [5], to achieve large-scale and long-range human perception. This sensor can intuitively provide the object distance information required for the robot to move without collision. Compared to traditional 2D laser rangefinders, it can provide laser measurement points in multiple planes, enabling the robot to detect objects based on a set of 3D points (called a point cloud) representing the environment. An intuitive understanding of the data generated by 3D lidar is shown in Fig. 1.2. Specifically, a tree search-based method is used to segment the point cloud, and then an Support Vector Machine (SVM) is used to classify the segments into two categories to determine which points represent humans and which points do not. This belongs to human detection. Subsequently, the segments representing different people in different point cloud frames are associated using the Global Nearest Neighbor (GNN) method, and their states are estimated using the Unscented Kalman Filter (UKF) method, to enable tracking of humans. The detection and tracking methods mentioned above are both statistical methods.

Regarding the second issue, this book focuses on a method that aims to give robots an **Online Learning (OL)** capability [6] (the idea is shown in Fig. 1.3), enabling them

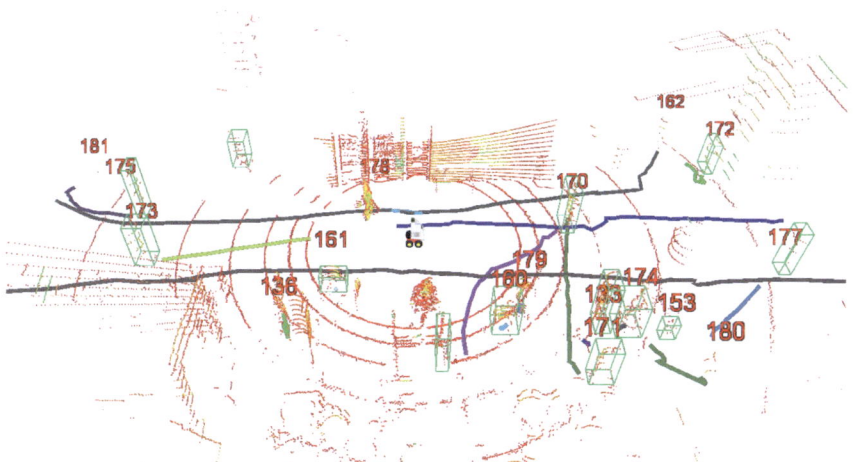

Fig. 1.2 Human detection and tracking in 3D point clouds [6]. Detected humans are enclosed in green bounding boxes. The colored lines are human movement trajectories generated by a multi-target tracker [7]

Fig. 1.3 Principle diagram of Robot Online Learning (ROL). It is easy to understand that by adding a learning module to the classic operational definition of service robots "sense-think-act", online learning is established, forming "sense-learn-think-act". The dashed boxes on both sides indicate that the modules included are closely related and can sometimes even be merged into one module

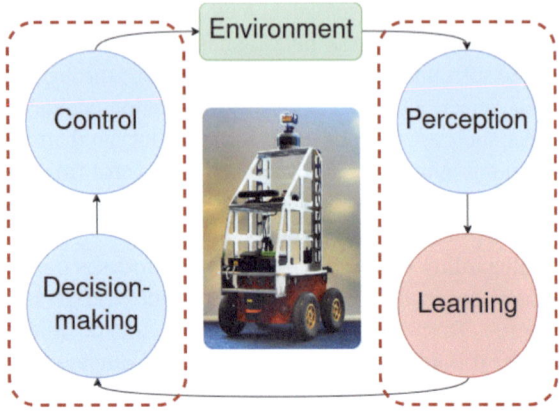

to absorb some new knowledge in the short term and maintain long-term memory of this knowledge. This kind of robot learning is on-site, on-the-fly, and faces a unique challenge that is different from the traditional field of machine learning: *the data entering the learning system is often unforeseen and unannotated.* It is worth pointing out that this book aims to propose a general learning method / framework (with dynamic representation capabilities) rather than a learning model. In other words, the proposed learning method should theoretically be able to incorporate different learning models. The theoretical basis of Robot Online Learning (ROL) is statistics.

It is worth mentioning that although this book focuses on the robot's learning of the environment, the robot's reasoning about the environment is highly related to the former and is often studied together. For example, a robot could learn to predict when humans are present at specific times and places by using, for example, statistical methods, including heat maps [8] and histograms [9–12], to build spatiotemporal models of the robot's long-term observations. On the other hand, the robot can also predict human motion trajectories, for example by using Long Short-Term Memory (LSTM) [13] to learn a predictive model from long-term robot deployment data [14].

The topological structure of the content of this book is shown in Fig. 1.4. Overall, the discussion is about how to build the embodied intelligence of robots from the two aspects of robot perception and learning. The downstream tasks mainly include human-aware robot navigation and long-term robot autonomy. In addition to the study of embodied intelligence methods themselves, the book also contains a chapter discussing how to effectively test and evaluate these methods and facilitate comparisons between different ones. The latter absorbs the agile development methods in the field of software engineering (as shown in Fig. 1.5), advocates the implementation of rapid closed-loop iterations for the development of new methods, and always prioritizes benchmarking when conducting experiments [15–19].

1.3 Related Open Source Projects

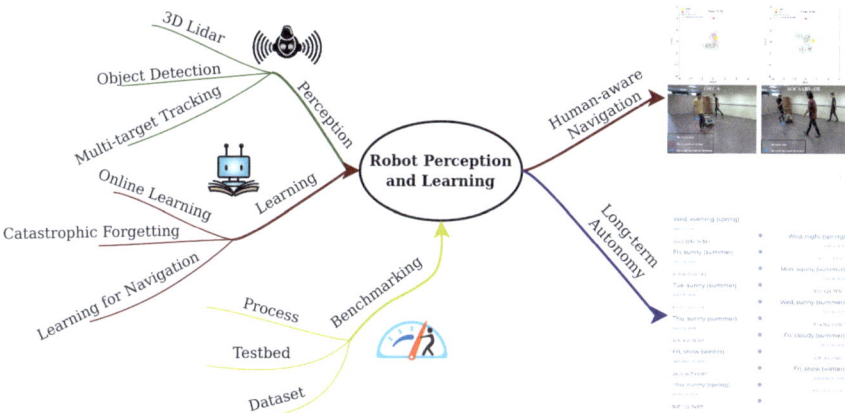

Fig. 1.4 Topology of the book's contents

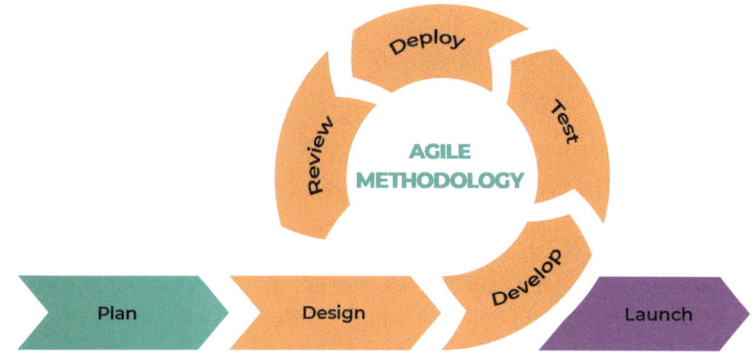

Fig. 1.5 Schematic diagram of agile development methodology in software engineering

1.3 Related Open Source Projects

Open science has been advocated and is gaining more and more attention. This section lists some open source projects related to the content of this book. Some of them will be mentioned again in their corresponding chapters.

- Open research data:
  - *L-CAS 3D Point Cloud People Dataset*[2] [6]: This dataset was collected with a 16-layer 3D lidar mounted on a mobile robot, in one of the main buildings of the University of Lincoln, UK. It includes 28,002 scan frames recorded by the robot while stationary and moving, with a total length of 49 min. About 20% of the data was manually annotated to form ground truth.

---

[2] https://lcas.lincoln.ac.uk/wp/research/data-sets-software/l-cas-3d-point-cloud-people-dataset/.

- *L-CAS Multisensor People Dataset*[3] [20]: This dataset is a supplement to the previous one, adding sensor data from an infrared camera and a 2D laser rangefinder.
- *FLOBOT Perception Dataset*[4] [8]: This dataset was collected using FLOBOT (an advanced autonomous floor scrubber) in public places in Italy and France including an airport, a supermarket and a warehouse. It includes data from four different sensors, including a 3D lidar and an RGB-D camera for human detection and tracking, and another RGB-D and a stereo camera for ground dirt and object detection. Additionally it contains the pose of the robot in the world reference frame.
- *EU Long-term Dataset with Multiple Sensors for Autonomous Driving*[5] [21]: This dataset was collected over a year in Montbéliard, France, in the city center (for long-term data) and in the suburbs (for roundabout data) using a vehicle equipped with 11 different sensors. For long-term data, the driving distance in each round is approximately 5.0 km (including a small loop and a large loop for loop closure), and the length of recorded data is approximately 16 min. For the roundabout data, the driving distance of each collection round is about 4.2 km (including 10 roundabouts of different sizes), and the recording data length is about 12 min.

- Open source code:

  - *Adaptive Clustering*[6] [6]: This is a lightweight and accurate point cloud clustering method implemented in C++.
  - *Online Learning for Human Classification in 3D LiDAR-based Tracking*[7] [6]: This is the code released with the paper, which allows the robot to learn a human model in the point cloud online at runtime without human intervention.
  - *Online Continual Learning for 3D Detection of Road Participants in Autonomous Driving*[8] [18, 19, 22]: This is the code released with the paper, which allows self-driving cars to learn models of various road participants from point clouds online at runtime, quickly, continuously, and without human intervention.
  - *Online Context Learning for Socially-compliant Navigation*[9] [23]: This is the code released with the paper, which allows mobile robots to discover different human social contexts when deployed in changing or across environments, thereby generating robot navigation with social attributes that conform to the context.

---

[3] https://lcas.lincoln.ac.uk/wp/research/data-sets-software/l-cas-multisensor-people-dataset/.
[4] http://lcas.github.io/FLOBOT/.
[5] https://epan-utbm.github.io/utbm_robocar_dataset/.
[6] https://github.com/yzrobot/adaptive_clustering.
[7] https://github.com/yzrobot/online_learning.
[8] https://github.com/RuiYang-1010/efficient_online_learning, https://github.com/RuiYang-1010/lstol.
[9] https://github.com/Nedzhaken/SOCSARL-OL.

- Open software and hardware:
  - *Multi-robot Exploration Testbed*[10] [24]: This testbed allows dozens of mobile robots to be repeatedly deployed in a 3D simulation scene driven by a physics engine to perform environment exploration tasks, and automatically collect various performance data during task execution for later analysis.
  - *L-CAS 3D Point Cloud Annotation Tool*[11] [25]: This tool provides semi-automatic annotation of point cloud data, whereby the point cloud is first automatically segmented and then each segment is labeled by humans.
  - *Human-aware Robot Navigation System*[12] [15, 26]: This is an open source hardware and software integration solution for building a mobile robot with human-aware navigation capabilities.

- Open educational resources:
  - *Introduction to Mobile Robotics*[13]: This course is designed to introduce basic concepts and techniques used in the field of mobile robotics. Relevant fundamental problems and challenges are analyzed, and both classic and cutting-edge solutions are illustrated.

1.4 Book Organization

Chapter 2 introduces the work on mobile robot software engineering, which mainly includes benchmarking of robot performance including evaluation methods, metrics, construction of testbeds and datasets, etc. In addition, some insights and thoughts on how to integrate AI into testing tools, benchmarking ethics, and data privacy, which are some of the aspects involved in modern AI development, is given. Although the latter are not the focus of the discussion in this chapter, it is still interesting to share some of the author's experiences and opinions, because these issues are increasingly unavoidable in today's research activities. Starting the main body of the book with the topic of benchmarking is a bit like the idea of "testing before development" in the field of software engineering, that is, we need to understand how to evaluate them before we really discuss the methods for realizing embodied intelligence. Moreover, this also happens to be consistent with the actions of the European Commission, that is, it would be a good choice to formulate a standard before vigorously developing AI to prevent its development from being uncontrolled and unregulated, which would lead to disastrous consequences [27].

---

[10] https://github.com/yzrobot/mrs_testbed.
[11] https://github.com/yzrobot/cloud_annotation_tool.
[12] https://github.com/Nedzhaken/human_aware_navigation.
[13] https://yzrobot.github.io/introduction_to_mobile_robotics/.

Chapter 3 introduces the work on robot perception. In a systematic way, the research motivation is first introduced, which is to use embodied sensors and onboard computing for large-scale human detection and tracking in public (non-home) environments. Then the contemporary 3D lidar adopted as an embodied sensor, from its basic working principle to its relevant applications in the field of mobile robots, is introduced. Next, the "adaptive clustering" method developed by the author is introduced, and the advantages and limitations of the proposed method is illustrated by comparing its performance with other popular methods at the time. Then, several hand-crafted features extracted from point clouds with proven performance for human model training are introduced. Finally, a multi-target tracker optimized for deployment in point clouds is introduced.

Chapter 4 introduces the work on robot learning, including a systematic study of ROL frameworks. First, what is ROL and why robots need OL are explained. Then, two ROL frameworks are introduced, one based on P-N learning and the other based on knowledge transfer. The advantages and disadvantages of the two methods are analyzed. Simply put, the former does not require external help but will produce self-bias, while the latter can avoid self-bias but requires external help. In addition, the latter also needs to resolve conflicts between internal and external parties. Next, the issue of how to alleviate catastrophic forgetting in the long-term process of ROL is addressed. Finally, how to leverage ROL to improve the performance of socially-compliant robot navigation in long-term and cross-environment deployments is introduced.

Chapter 5 summarizes the full text and gives prospects for future research.

References

1. Turing, A.M.: Computing machinery and intelligence. Mind **59**(236), 433–460 (1950). https://doi.org/10.1093/mind/LIX.236.433
2. Yan, Z., Jouandeau, N., Ali Cherif, A.: A survey and analysis of multi-robot coordination. Int. J. Adv. Robot. Syst. **10**(399), 1–18 (2013)
3. Bellotto, N., Cosar, S., Yan, Z.: Human detection and tracking. In: Ang, M.H., Khatib, O., Siciliano, B. (eds.) Encyclopedia of Robotics, pp. 1–10. Springer (2018)
4. Yan, Z., Sun, L., Krajnik, T., Duckett, T., Bellotto, N.: Towards long-term autonomy: a perspective from robot learning. In: Proceedings of the AAAI-23 Bridge Program on AI and Robotics, pp. 1–4. Washington, USA (2023)
5. Yang, T., Li, Y., Zhao, C., Yao, D., Chen, G., Sun, L., Krajnik, T., Yan, Z.: 3D ToF LiDAR in mobile robotics: a review. CoRR abs/2202.11025 (2022). URL http://arxiv.org/abs/2202.11025
6. Yan, Z., Duckett, T., Bellotto, N.: Online learning for human classification in 3D LiDAR-based tracking. In: Proceedings of the 2017 IEEE/RSJ International Conference on Intelligent Robots and Systems (IROS), pp. 864–871. Vancouver, Canada (2017)
7. Bellotto, N., Hu, H.: Computationally efficient solutions for tracking people with a mobile robot: an experimental evaluation of Bayesian filters. Autonom. Robots **28**(4), 425–438 (2010). https://doi.org/10.1007/S10514-009-9167-2
8. Yan, Z., Schreiberhuber, S., Halmetschlager, G., Duckett, T., Vincze, M., Bellotto, N.: Robot perception of static and dynamic objects with an autonomous floor scrubber. Intell. Serv. Robot. **13**(3), 403–417 (2020)

References

9. Vintr, T., Eyisoy, K., Vintrova, V., Yan, Z., Ruichek, Y., Krajnik, T.: Spatiotemporal models of human activity for robotic patrolling. In: Proceedings of the 6th International Conference on Modelling and Simulation for Autonomous System (MESAS), pp. 54–64. Prague, Czech Republic (2018)
10. Vintr, T., Krajnik, T., Molina, S., Senanayake, R., Broughton, G., Yan, Z., Ulrich, J., Kucner, T., Swaminathan, C., Majer, F., Stachova, M., Lilienthal, A.: Time-varying pedestrian flow models for service robots. In: Proceedings of the 2019 European Conference on Mobile Robots (ECMR), pp. 1–7. Prague, Czech Republic (2019)
11. Vintr, T., Molina, S., Senanayake, R., Broughton, G., Yan, Z., Ulrich, J., Kucner, T.P., Swaminathan, C.S., Majer, F., Stachova, M., Lilienthal, A.J., Krajnık, T.: Spatio-temporal representation of time-varying pedestrian flows. In: Proceedings of the ICRA Workshop on Long-Term Human Motion Prediction, pp. 1–2. Montreal, Canada (2019)
12. Vintr, T., Yan, Z., Duckett, T., Krajnik, T.: Spatio-temporal representation for long-term anticipation of human presence in service robotics. In: Proceedings of the 2019 IEEE International Conference on Robotics and Automation (ICRA), pp. 2620–2626. Montreal, Canada (2019)
13. Sherstinsky, A.: Fundamentals of recurrent neural network (RNN) and long short-term memory (LSTM) network. Phys. D Nonlinear Phenom. **404**, 132306 (2020)
14. Sun, L., Yan, Z., Mellado, S.M., Hanheide, M., Duckett, T.: 3DOF pedestrian trajectory prediction learned from long-term autonomous mobile robot deployment data. In: Proceedings of the 2018 IEEE International Conference on Robotics and Automation (ICRA), pp. 5942–5948. Brisbane, Australia (2018)
15. Okunevich, I., Hilaire, V., Galland, S., Lamotte, O., Shilova, L., Ruichek, Y., Yan, Z.: Human-centered benchmarking for socially-compliant robot navigation. In: Proceedings of the 2023 European Conference on Mobile Robots (ECMR), pp. 1–7. Coimbra, Portugal (2023)
16. Vintr, T., Blaha, J., Rektoris, M., Ulrich, J., Roucek, T., Broughton, G., Yan, Z., Krajnik, T.: Toward benchmarking of long-term Spatio-temporal maps of pedestrian flows for human-aware navigation. Front. Robot. AI **9**, 890013 (2022)
17. Vintr, T., Yan, Z., Eyisoy, K., Kubis, F., Blaha, J., Ulrich, J., Swaminathan, C.S., Mellado, S.M., Kucner, T., Magnusson, M., Cielniak, G., Faigl, J., Duckett, T., Lilienthal, A.J., Krajnik, T.: Natural criteria for comparison of pedestrian flow forecasting models. In: Proceedings of the 2020 IEEE/RSJ International Conference on Intelligent Robots and Systems (IROS), pp. 11197–11204. Las Vegas, USA (2020)
18. Yang, R., Yan, Z., Yang, T., Ruichek, Y.: Efficient online transfer learning for 3D object classification in autonomous driving. In: Proceedings of the 2021 IEEE International Conference on Intelligent Transportation Systems (ITSC), pp. 2950–2957. Indianapolis, USA (2021)
19. Yang, R., Yan, Z., Yang, T., Wang, Y., Ruichek, Y.: Efficient online transfer learning for road participants detection in autonomous driving. IEEE Sens. J. **23**(19), 23522–23535 (2023)
20. Yan, Z., Sun, L., Duckett, T., Bellotto, N.: Multisensor online transfer learning for 3D lidar-based human detection with a mobile robot. In: Proceedings of the 2018 IEEE/RSJ International Conference on Intelligent Robots and Systems (IROS), pp. 7635–7640. Madrid, Spain (2018)
21. Yan, Z., Sun, L., Krajnik, T., Ruichek, Y.: EU long-term dataset with multiple sensors for autonomous driving. In: Proceedings of the 2020 IEEE/RSJ International Conference on Intelligent Robots and Systems (IROS), pp. 10697–10704. Las Vegas, USA (2020)
22. Yang, R., Yang, T., Yan, Z., Krajnik, T., Ruichek, Y.: Preventing catastrophic forgetting in continuous online learning for autonomous driving. In: Proceedings of the 2024 IEEE/RSJ International Conference on Intelligent Robots and Systems (IROS), pp. 1–8. Abu Dhabi, UAE (2024)
23. Okunevich, I., Lombard, A., Krajnik, T., Ruichek, Y., Yan, Z.: Online context learning for socially compliant navigation. IEEE Robot. Autom. Lette. Early Access **14**, 1–8 (2025)
24. Yan, Z., Fabresse, L., Laval, J., Bouraqadi, N.: Building a ros-based testbed for realistic multi-robot simulation: taking the exploration as an example. Robotics **6**(3), 1–21 (2017)
25. Yan, Z., Duckett, T., Bellotto, N.: Online learning for 3D lidar-based human detection: experimental analysis of point cloud clustering and classification methods. Autonom. Robots **44**(2), 147–164 (2020)

26. Okunevich, I., Hilaire, V., Galland, S., Lamotte, O., Ruichek, Y., Yan, Z.: An open-source software-hardware integration scheme for embodied human perception in service robotics. In: Proceedings of the 2021 IEEE International Conference on Advanced Robotics and Its Social Impacts (ARSO), pp. 1–6. Hong Kong, China (2024)
27. Regulation (eu) 2024/1689 of the European parliament and of the council of 13 June 2024 laying down harmonised rules on artificial intelligence and amending regulations (ec) no 300/2008, (eu) no 167/2013, (eu) no 168/2013, (eu) 2018/858, (eu) 2018/1139 and (eu) 2019/2144 and directives 2014/90/eu, (eu) 2016/797 and (eu) 2020/1828

Chapter 2
Benchmarking in Mobile Robotics

Abstract This chapter presents a study on software engineering for mobile robotics, focusing on benchmarking robot performance, including evaluation methods, metrics, and the construction of testbeds and datasets. It also offers insights into integrating AI into testing tools, benchmarking ethics, and data privacy, aspects increasingly relevant to modern AI development. While these latter topics are not the primary focus of this chapter, sharing the author's experiences and perspectives is nonetheless valuable, given their growing importance in contemporary research. Beginning this book with benchmarking aligns with the software engineering principle of "testing before development", emphasizing the need to establish evaluation methodologies before discussing methods for realizing embodied intelligence. This approach also resonates with the European Commission's emphasis on formulating standards prior to extensive AI development to mitigate the risks of uncontrolled and unregulated advancements, which could lead to detrimental consequences.

Keywords Benchmarking · Evaluation metrics · Testbed · Dataset

2.1 Introduction

Benchmarking refers to the evaluation of different methods using the same evaluation process—including using the same test data—and the same evaluation metrics. From author's point of view, benchmarking is a double-edged sword. On the one hand it can make comparisons between different methods easier (e.g. without reproducing the results of the compared methods) and fairer, but on the other hand it drives methods to overfit to specific benchmarks. Regarding the former aspect, a pioneering work in the robotics community is the benchmarking of different Simultaneous Localization And Mapping (SLAM) methods [1]. Other work includes the KITTI [2] and Waymo [3] benchmark suites, which are widely used in modern autonomous driving. Regarding the latter aspect, further explanation is given at the end of this chapter using the KITTI benchmark suite as an example.

2.2 Benchmarking Process

The benchmarking process broadly consists of four parts/phases. As shown in Fig. 2.1, first, the experimental design and execution rules are clearly defined by the experimenter. The former includes the specified experimental parameters and the data that need to be collected during the experiment run. The latter includes the specific steps to conduct the experiment, the number of repetitions, completion conditions, results submission methods, and so forth. Then there is the execution of the experiment, which is ideally fully automated (i.e. without human intervention) in order to increase experimental efficiency. Unless the experiment is deterministic, i.e. does not contain any stochastic component or model, it should be repeated multiple times so that the results can be subsequently analyzed statistically. Furthermore, repeated runs of the same experiment should be independent of each other and can be performed sequentially or in parallel depending on available resources. Later, the collected experimental data is evaluated using defined metrics. Finally, the evaluation results of different methods following the same process are organized to form benchmark results in order to characterize their advantages and disadvantages.

2.2.1 Parameters

Benchmark parameters need to be task-specific. Effectively identifying and defining them is critical to benchmarking as well as the development and improvement of methods. Operational Design Domain (ODD), which has attracted attention in the field of autonomous driving in recent years, is a typical example with the characteristics of the times. It aims to detail all conceivable overlapping conditions, use cases, restrictions and scenarios that autonomous vehicles might encounter, even the most unusual (edge) cases. Common ODD conditions include lighting, weather, terrain, and road type, but enumerating them all is very challenging. Three specific examples are used below to further illustrate the parameters in benchmarking.

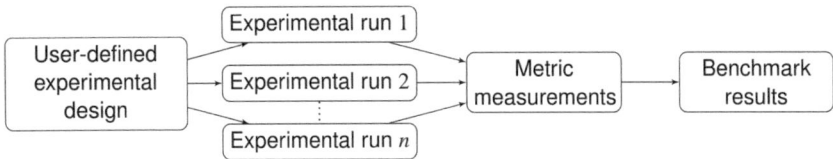

Fig. 2.1 A full benchmarking process: from experimental design to metric measurements [4]

2.2 Benchmarking Process

2.2.1.1 Multi-Robot Exploration

Due to the complexity of multi-robot systems, there are many parameters that can influence the experimental results. The most relevant parameters for multi-robot systems to perform exploration tasks are explored here as per [4], including three aspects: "robot", "team" and "environment". The goal is to provide the community with a basis for building a database of benchmark settings, where each setting corresponds to a different set of parameter configurations.

- Robot:
    - *Geometric characteristics* such as size and shape.
    - *Physical characteristics* such as weight and appearance.
    - *Chemical characteristics* such as material (involving sensor (installed on other robots) or environmental sensitivity).
    - *Mechanical characteristics* such as holonomic or non-holonomic.
    - *Kinematic characteristics* such as speed and acceleration.
    - *Integrated sensor properties* such as sensing modality and sampling frequency.
    - *Onboard computing resources* such as CPU and RAM.
- Team:
    - *Number of robots*. The greater the number does not mean the better the performance of the multi-robot system [5].
    - *Qualitative nature of the team* such as homogeneous or heterogeneous.
    - *Initial position of the robots* which may have a significant impact on the team's exploration performance [5].
    - *Means of communication* such as explicit communication or implicit communication [6]. The former further includes parameters such as communication bandwidth and communication range, while the latter further includes parameters such as communication media and information life cycle.
- Environment:
    - *Area*. Exploring a large environment may be more challenging than exploring a small one (due to sensor capabilities, robot endurance, the robot's or team's ability to handle accumulated errors, etc.)
    - *Terrain* such as flat ground and slope.
    - *Landform* include the form and layout of the environment, the shape and density of obstacles, etc.
    - *Material of obstacles*. For example, lidar cannot easily detect glass but sonar can, too many walls can affect Wi-Fi signal, etc.
    - *Surface texture* such as grass and gravel.
    - *Dynamics* such as slow changes like layout and fast changes like pedestrians.
    - *Weather condition* such as sunny or adverse.
    - *Lighting condition* which mainly affects passive visual sensors.

2.2.1.2 Detection of Road Participants

The parameters that affect the performance of road participant detection [7, 8], or more generally, object detection based on sensor data, can also be divided into three categories including "robot", "object" and "environment". Overall, the identified parameters related to "robot" and "environment" are less than those for the "multi-robot exploration" task, since the reactive perception of road participants is the focus of this book. In contrast, if robots or autonomous vehicles perform detection tasks in a proactive manner, then the same parameters defined in the "multi-robot exploration" task can be shared.

- Robot:
  - *Integrated sensor properties* which constitute decisive parameters, since robot perception relies on various sensors [9].
  - *Onboard computing resources*. In addition to traditional CPU and RAM, GPU has become one of the most important parameters in this era.
- Object:
  - *Geometric characteristics* such as size and shape.
  - *Physical characteristics* such as color and texture.
  - *Chemical characteristics* such as surface material (fabric, metal, rubber, etc.)
  - *Biological characteristics* such as temperature and movement patterns (e.g. humans walking upright [10]).
  - *Kinematic characteristics* such as speed and acceleration.
- Environment:
  - *Dynamics* which mainly cause occlusion of objects and affect context-based detection.
  - *Material of occluders*. For example, some millimeter wave radars can penetrate drywall, wood, glass and more [11].
  - *Weather condition* such as sunny or adverse.
  - *Lighting condition* which mainly affects passive visual sensors.
  - *Temperature condition* which can affect certain temperature-sensitive sensors such as thermal imagers [12]. Or, too low or too high ambient temperature can cause the camera to generate more noisy data.
  - *Discriminability* represents the extent to which the robot is able to distinguish objects from the environment.

2.2.1.3 Human-Aware Robot Navigation

This task [13] can be seen as a cross between the previous two, taking into account the parameters of both the robot as an active entity and the human being as the

2.2 Benchmarking Process

detected object. In addition, the interaction between robots and humans needs to be considered, which in this book aims to be socially-compliant. The latter prompts the need to additionally discern two parameters in the navigation task, each from the "robot" and "human" aspects.

- Robot: In addition to common parameters with the above two tasks including geometric, physical, mechanical and kinematic characteristics, integrated sensor properties, and onboard computing resources, it also includes:

  – *Sociality* such as pleasing or repulsive appearance.

- Human: In addition to the same parameters related to the "object" as in the previous task, including physical, chemical, biological and kinematic characteristics, it also includes:

  – *Psychological factors* such as acceptance and trust in robots.

- Environment: This part of the parameters takes the set of corresponding parameters of the above two tasks.

2.2.2 Metrics

Metrics, or sometimes Key Performance Indicators (KPIs), are used to quantify the performance of a system (or method, model, etc.) in order to analyze its strengths and weaknesses and also to facilitate comparisons between different methods. Metrics can be task-independent or task-related. An example of the former is the performance evaluation of the robot itself, such as onboard computing power, which can be achieved by quantifying the performance of integrated components such as CPU, RAM, network transmission, etc. The definition of task-related metrics aims to evaluate the ability of robots to complete specific tasks, and it should strive to be standardizable. Below, corresponding to the three examples mentioned in Sect. 2.2.1, the evaluation metrics used are presented.

2.2.2.1 Multi-Robot Exploration

For the multi-robot exploration task, the overall performance of the robot team rather than the performance of a single robot is evaluated because, essentially, it is hoped that the performance differences between different multi-robot coordination methods can be reflected through evaluation from a global perspective.

Exploration time

This is one of the most commonly used metrics for exploration tasks and measures the time it takes for a team of robots to complete a given exploration task [5, 14, 15]. Its definition can be rigorously described as: timing starts when any robot in the team starts to perform the exploration task and ends when any robot in the team obtains a predetermined percentage of the exploration information (such as a map) in the specified space. Time is measured in wall clock time. Moreover, one of the objectives of multi-robot exploration optimization is to minimize the overall exploration time. The challenge in achieving this remains to move each robot to an optimal position that maximizes the exploration area (i.e. information gain) and simultaneously minimizes robot usage (e.g., the "exploration cost" mentioned below). Unfortunately, this problem is known to be NP-hard.

Exploration cost

The distance traveled is often used to estimate the cost of a mobile robot performing a task [5]. The exploration cost is defined as the sum of the distances traveled by all robots in a team:

$$\text{cost}(n) = \sum_{i=1}^{n} d_i \qquad (2.1)$$

where n is the number of robots in the team, d_i is the distance traveled by robot i. In fact, in multi-robot exploration an estimate of the distance each robot will travel is often used to calculate the cost for task allocation [14–16]. Furthermore, the exploration cost can have different definitions according to user needs, such as energy consumption, computing and communication resource occupation, etc.

Exploration efficiency

Efficiency is usually measured as the ratio of useful output to total input. In terms of exploration, it is directly proportional to the amount of information the robot team collects from the environment and inversely proportional to the cost it incurs [17]:

$$\text{efficiency}(n) = \frac{A}{\text{cost}(n)} \qquad (2.2)$$

where A is the total area explored. For example, the area is measured in square meters, and if the exploration efficiency is 1.6, this means that, on average, every time the sum of all robot movements in a team is 1 m, the entire team should have explored an area of 1.6 m^2.

2.2 Benchmarking Process

Exploration safety

Collision avoidance is one of the basic requirements for mobile robots. For multi-robot systems, the risk of collision increases with the size of the team. Therefore, the safety metric valued in robotic systems is defined as:

$$\text{safety}(n) = 1 - \frac{\sum_{i=1}^{n} s_i}{S} \qquad (2.3)$$

where S represents a predefined base, s_i is the number of collisions experienced by robot i. The larger the value of safety, the higher the safety of the multi-robot system being evaluated.

2.2.2.2 Detection of Road Participants

The object detection task is generally defined as given a frame of sensor data, finding the object of interest and determining its location and category (or class). Measuring its performance requires introducing a concept called "ground truth". The latter refers to the true label or reference data used to train and evaluate supervised learning models, which is usually provided by humans. However, it is important to note that the ground truth is sometimes not objective reality, but rather a human interpretation, and thus may contain biases and errors. Calculating the various differences between the results given by the computer and the ground truth constitutes the evaluation of object detection performance. The metrics to be introduced in this section include confusion matrix, F-score, Intersection over Union (IoU) and Average Precision (AP). The first two are often used to evaluate object classification performance, while the latter two are used to evaluate object detection performance.

Confusion matrix

The confusion matrix is a specific tabular data that allows a visual display of the classification results of a model. It is particularly suitable for multi-classification problems, as it not only shows the classification performance but also shows which class was incorrectly classified. Each row of the matrix represents an instance in the actual class, while each column represents an instance in the predicted class. For an intuitive understanding, Fig. 2.2 shows a confusion matrix for a binary classification problem. Where:

- True Positive (TP) means a positive example is predicted correctly.
- False Positive (FP) means a negative example is predicted incorrectly.
- False Negative (FN) means a positive example is predicted incorrectly.
- True Negative (TN) means a negative example is predicted correctly.

F-score

The F-score, also known as the F1-score or F-measure, is a commonly used metric used to evaluate the performance of a binary classification model. It takes into account the precision and recall of the algorithm, aiming to find a balance between the two. Precision is the proportion of positive results that are actually positive. Recall is the proportion of actual positive cases that are correctly identified. The F-score is the harmonic mean of precision and recall, calculated as follows:

$$F = 2 \cdot \frac{\text{precision} \cdot \text{recall}}{\text{precision} + \text{recall}} \quad (2.4)$$

where

$$\text{precision} = \frac{TP}{TP + FP}, \quad \text{recall} = \frac{TP}{TP + FN} \quad (2.5)$$

In multi-classification problems, there are two ways to calculate F-score: one is called Micro F-score and the other is called Macro F-score. The former is calculated by first summing the true positives and false positives for all classes, and then calculating the F-score from these totals. In other words, it treats all classes as if they were one big class. While the latter is calculated by first calculating the F-score for each class, and then averaging the scores across all classes. In other words, it treats each class as its own entity. The advantage of Micro F-score is that it can reflect the overall performance across all classes, but the disadvantage is that it might overestimate performance due to the dominance of the majority class in imbalanced datasets. The advantage of Macro F-score is that it gives equal weight to each class, ensuring all classes contribute equally to the score, but the disadvantage is that it might not directly reflect the overall performance, especially if class distributions differ signif-

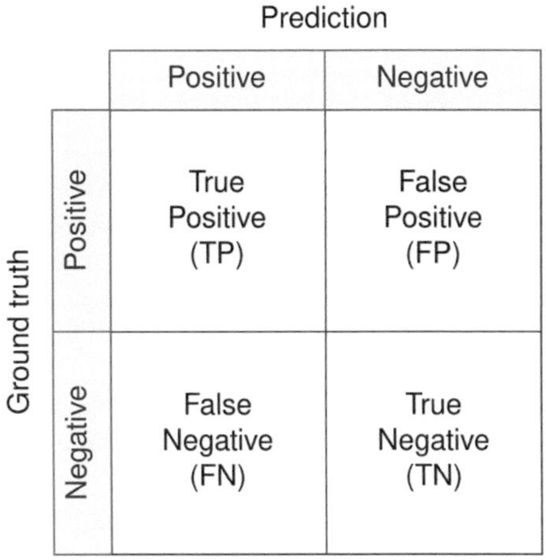

Fig. 2.2 Confusion matrix for binary classification

		Prediction	
		Positive	Negative
Ground truth	Positive	True Positive (TP)	False Positive (FP)
	Negative	False Negative (FN)	True Negative (TN)

2.2 Benchmarking Process

icantly. For example, the overall performance is good, but some individual classes have low F-scores. In summary, it's often beneficial to consider both Micro F-score and Macro F-score to gain a comprehensive understanding of a model's performance across different aspects.

Intersection Over Union

IoU, also known as the Jaccard index, calculates the ratio of the intersection and union of the "predicted bounding box" and the "true bounding box" (i.e. ground truth), which is formulated as follows:

$$\text{IoU}(B_p, B_t) = \frac{\text{Area}(B_p \cap B_t)}{\text{Area}(B_p \cup B_t)} \tag{2.6}$$

which can be visualized as [18]:

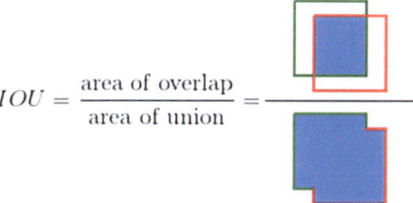

$$IOU = \frac{\text{area of overlap}}{\text{area of union}} =$$

The above definition and calculation consider a 2D plane, but can be straightforwardly extended to 3D space. In general, the IoU threshold is set to 0.5, that is, when IoU ≥ 0.5, it is considered a true detection, otherwise it is considered a false detection. However, the community usually adjusts the corresponding IoU according to the size of the objects being detected. For example, the KITTI [2] and Waymo [3] evaluation suites employ a 50% overlap threshold for pedestrians and cyclists but a 70% threshold for cars.

Average Precision

Similar to the F-score, AP also takes into account both the precision and recall of a model, but measures the area under the precision-recall curve, which represents the overall ability of the model to rank positive examples above negative ones. A higher AP value indicates a better balance between precision and recall, meaning the model can correctly identify positives while minimizing false positives. It is formalized as follows:

$$\text{AP} = \int_0^1 p(r)dr \tag{2.7}$$

where $p(r)$ is the precision at recall level r.

In actual calculation, the interpolation method can be used:

$$\text{AP} = \frac{1}{|R|} \sum_{r \in R} p_{interp}(r) \tag{2.8}$$

where $p_{interp}(r) = \max_{r':r' \geq r} p(r')$, which means selecting the maximum recall rate at a fixed precision value. Taking the KITTI evaluation setting as an example, 11 equally spaced recall levels are applied, i.e., $R_{11} = \{0, 0.1, 0.2, \ldots, 1\}$. The limitation of AP is that its results are sensitive to the choice of IoU threshold and cannot distinguish the location accuracy of the predicted bounding box.

2.2.2.3 Socially-Compliant Robot Navigation

The social acceptability of robot navigation around humans can be evaluated from both human and robot perspectives. As human-aware robot navigation constitutes a form of Human-Robot Interaction (HRI), considering the experiences of both humans and robots leads to a more comprehensive and unbiased evaluation. Beyond the separate application of established Robot-Centric Metrics (RCM) and Human-Centric Metrics (HCM), recent research has explored the latent correlations between these metrics through extensive benchmarking [13]. This approach aims, first, to assess the social attributes of robot navigation using only RCM when HCM are impractical to apply, and second, to enable online optimization of the robot's social navigation performance based on real-time system evaluations derived from RCM [19].

Robot-centric metrics

Five RCMs commonly employed in the literature are listed below.

- The *extra time ratio* quantifies the additional time a robot requires to complete a task in a human-shared environment [20–22]. It is defined as:

$$R_{\text{time}} = \frac{T}{T_{\text{human}}} \tag{2.9}$$

where T and T_{human} are the time it takes for the robot to complete the task without and in the presence of humans, respectively.
- The *extra distance ratio* evaluates system performance by quantifying the additional distance a robot travels in the presence of humans [23, 24]. It is defined as:

$$R_{\text{distance}} = \frac{D}{D_{\text{human}}} \tag{2.10}$$

2.2 Benchmarking Process

where D and D_{human} denote the distance the robot needs to travel to complete a task without and in the presence of humans, respectively.

- The *success ratio* quantifies a robot's ability to complete a task without colliding with a human [20–22]. It is defined as:

$$R_{\text{success}} = \frac{N_{\text{success}}}{N} \quad (2.11)$$

where N_{success} denotes the number of successful trials in which the robot did not hit a human, and N represents the total number of trials.

- The *hazard ratio* quantifies the proportion of time a robot spends within a hazardous proximity of humans [22]. It is defined as:

$$R_{\text{hazard}} = \frac{1}{N} \sum_{i=1}^{N} \frac{T_i^{\text{hazard}}}{T_i^{\text{social}}} \quad (2.12)$$

where N is the number of humans present, T_i^{hazard} is the duration for which the distance between the robot and the i-th human is less than a predefined safe distance (e.g., 1 m), and T_i^{social} is the duration for which the distance between the robot and the i-th human is less than a predefined social distance (e.g., 1.5 m).

- The *deceleration ratio* quantifies a robot's ability to reduce its speed when approaching a human [25]. It is defined as:

$$R_{\text{deceleration}} = \frac{1}{N} \sum_{i=1}^{N} \frac{V_i}{V_{\text{max}}} \quad (2.13)$$

where N is the number of speed measurements taken when the distance between the robot and the human is less than a predefined social distance D_{social}, V_i is the robot's instantaneous speed at the i-th measurement, and V_{max} is the robot's maximum speed. Because different robots or methods may employ varying maximum speeds, comparisons using this metric can be ambiguous, making it difficult to isolate the impact of hardware versus algorithmic performance. Therefore, it is recommended to maintain a consistent maximum speed across different methods when using $R_{\text{deceleration}}$ for benchmarking.

Human-centric metrics

Essentially, RCM alone cannot fully account for the degree of sociality of robot navigation because they do not reflect humans' subjective feelings about robot behavior. To obtain people's opinions and quantify them, a questionnaire[1] based on the Robotic Social Attributes Scale (RoSAS) [26] can be used. RoSAS has been widely used to

---

[1] https://forms.gle/4Lr4KP1E81SJFAET9.

Table 2.1 The robotic social attributes scale (RoSAS)

Warmth	Competence	Discomfort
Happy	Capable	Scary
Feeling	Responsive	Strange
Social	Interactive	Awkward
Organic	Reliable	Dangerous
Compassionate	Competent	Awful
Emotional	Knowledgeable	Aggressive

assess human acceptance of robots in close-proximity HRI, such as object handover between humans and robots [27]. It was subsequently used to evaluate the sociality of robot navigation methods [13], since human-aware robot navigation can be viewed as a contactless HRI problem. The RoSAS questionnaire provides a psychologically validated and standardized evaluation scale for the quality of HRI. These questions are classified into three categories corresponding to three psychological factors including "warmth", "competence" and "discomfort", as shown in Table 2.1.

Furthermore, a counterpart (dual metric) to R_{time} within the realm of HCM can be defined: the *human extra time ratio*. This metric aims to quantify how human task completion time is affected by the presence of robots exhibiting varying degrees of social behavior. It is objectively calculable and defined as:

$$R'_{\text{time}} = \frac{T'}{T'_{\text{robot}}} \tag{2.14}$$

where T' and T'_{robot} represent the time a human takes to complete a task in the absence and presence of robots, respectively.

2.2.3 Experimental Design

Scientific and reasonable experimental design is the foundation of benchmarking. Experimental design and its associated terminology vary across disciplines. Based on the author's research experience [4, 13, 28–31], it is recommended to consider the following aspects when designing robotic experiments:

- *Experimental environment*: Specify whether experiments are conducted in simulation, using a dataset, or with a physical robot.
- *Experimental subject*: Define the focus of the experiment: a specific module, algorithm, or the entire system.
- *Experimental scope*: Characterize the scope as end-to-end (macroscopic, akin to black-box testing in software engineering) or structured (microscopic, akin to white-box testing).

2.2 Benchmarking Process

- *Experimental parameters*: For each parameter, provide the specific values or range of values used in the experiments. Examples are detailed in Sect. 2.2.1.
- *Parameter vectors*: Define a parameter vector as the set of parameters and their corresponding values for a single experimental run. Describe how these vectors are generated (e.g., explicit listing or computational method).
- *Experiment repetitions*: Specify the number of repetitions for each experimental configuration. Justify the choice of repetitions, explaining how it addresses repeatability and enables statistical analysis, particularly for non-deterministic experiments common in mobile robotics due to factors such as sensor noise or probabilistic algorithms.
- *Experiment termination criteria*: Define the conditions under which an experimental run is terminated, whether upon successful completion or due to unexpected behavior.
- *Data acquisition plan*: Describe the data collection strategy, including when and how data are acquired and stored. For example, robot data might be logged continuously at a defined frequency, while human feedback might be collected via post-experiment questionnaires.
- *Data acquisition (Measurements)*: Specify the types of data collected for each experimental run and the corresponding acquisition methods. Data collection may be continuous (at a predetermined frequency) or discrete (triggered by specific events or conditions).
- *Evaluation metrics*: Define the metrics used to evaluate performance based on the collected data. See Sect. 2.2.2 for further details.
- *Data analysis and results presentation*: Describe the data processing procedures used to analyze the collected data (e.g., to understand data structure and distribution) and present the experimental results. This may include visualizations (e.g., precision-recall curves) and comparative rankings (e.g., as in the KITTI benchmark suite).

To illustrate the aforementioned arguments, an example experimental design for a multi-robot exploration task is presented, as shown in Table 2.2. This design, originally published in [4], is presented here in an expanded form. Creating such a table for each benchmark, detailing all aspects of the experiment, facilitates both *reproducibility* and *replicability*. Reproducibility focuses on the consistency of experimental conclusions. For instance, if experimenter B conducts the experiment in a different environment and arrives at the same conclusion as experimenter A (e.g., "Algorithm I is faster than Algorithm II"), the experiment is considered reproducible. Replicability, conversely, emphasizes the consistency of experimental results, that is, obtaining nearly identical results upon repetition (e.g., "Algorithm I is 100 seconds faster than Algorithm II"). Achieving replicability is often more challenging than reproducibility, as it typically requires identical benchmarking platforms, including hardware and software configurations.

Table 2.2 Example experimental design for multi-robot exploration

Experimental environment	Simulator: MORSE [32]
Experimental subject	Map merging algorithm: probabilistic merging [14]
Experimental scope	End-to-end and structured
Experimental parameters	Robot: Pioneer 3-DX, Number of robots: [3, 33], $\cdots$
Parameter vectors	All possible combinations
Experiment repetitions	5 times
Experiment termination criteria	99% of the area is explored, the experiment run >10 mins, $\cdots$
Data acquisition plan	The robot collects data at runtime and stores it locally
Data acquisition	Area explored per simulation step, exploration time when a stop condition is triggered, $\cdots$
Evaluation metrics	Exploration time, map quality [34], $\cdots$
Data analysis and results presentation	Analyze raw data statistically, visualize results using line graphs, $\cdots$

2.3 Testbed Construction

Benchmarking requires a platform, which can be implemented using a real robot, a simulator, a dataset, or a combination thereof. As an example, this section details a simulation testbed developed for benchmarking multi-robot exploration tasks [4]. Methodologically, the testbed employs the physics engine-based simulator MORSE [32] as the simulation front end, the Robot Operating System (ROS) [35] middleware as the robot software interface, a computer cluster for large-scale simulations (back end), and a monitoring system to oversee experiment execution. From an engineering standpoint, the testbed fully automates the benchmarking process, generating evaluation results without human intervention. The testbed's architecture is illustrated in Fig. 2.3.

The MORSE simulator and the experiment monitor are deployed as independent ROS nodes on a workstation equipped with an 8-core processor, 8 GB of RAM, a high-performance graphics card, and a Gigabit Ethernet adapter. The robot controllers are deployed on the computer cluster, with each controller implemented as a set of ROS nodes. To ensure fidelity between simulation and real-world experiments, the same controller software used on the physical robot is employed in the simulation. The computer cluster comprises 70 computing nodes, providing high-performance distributed computing resources to support real-time simulation of large robot teams. Each node is equipped with 8–12 processors and 16–48 GB of RAM. Wired networks facilitate communication within the cluster and between the cluster

2.4 Dataset Building

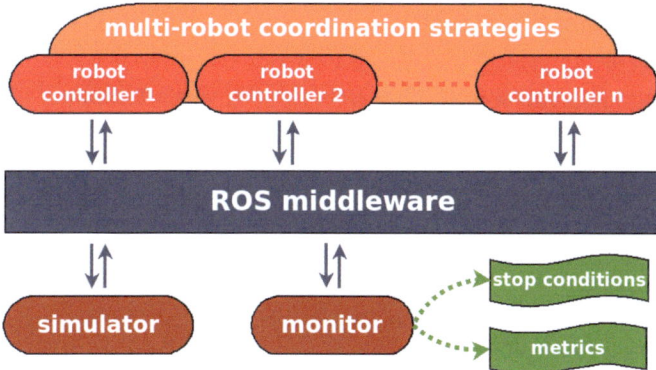

Fig. 2.3 Architecture of the developed multi-robot simulation testbed

and the workstation. Inter-node communication within the ROS framework is based on the publish-subscribe model.

Communication within the testbed is threefold:

- Simulator-controller communication: The simulator transmits simulated sensor data (e.g., laser rangefinder and wheel encoder data) to the robot controllers, which, in turn, send motion control commands to the simulator.
- Monitor-controller communication: The monitor receives performance metrics from the robot controllers, such as explored map coverage and distance traveled.
- Inter-robot communication: Communication between robot controllers depends on the specific coordination strategy employed, and may include exchanging map and localization information.

In general, the data volume and network bandwidth requirements scale proportionally with the number of simulated robots.

2.4 Dataset Building

As another example of a benchmarking platform, this section introduces the EU long-term dataset [9], designed for evaluating robot perception and learning methods. This dataset was acquired using an embodied perception system specifically developed for autonomous driving. A key design principle of this system is its multimodal approach to perception, contrasting with Tesla's current unimodal (camera-based) approach. Detailed characteristics and analyses of the dataset are available on its website. The following provides a detailed description of the dataset's construction from both hardware and software perspectives.

2.4.1 Hardware Platform

The hardware platform is the UTBM RoboCar [9], which features a more diverse sensor suite compared to platforms used in previous works, such as the KITTI AnnieWAY [2] and Oxford RobotCar [36]. This enhanced diversity includes the integration of more and different sensors. Furthermore, a key design principle for the selection and installation of exteroceptive sensors was to maximize perceptual coverage and ensure sensor redundancy (i.e., overlapping sensor field of view (FoV), as illustrated in Fig. 2.4).

The sensor types and their mounting positions on the vehicle are visualized in Fig. 2.5. Specifically:

- *Stereo cameras*: Two stereo camera pairs, a forward-facing Bumblebee XB3 and a rear-facing Bumblebee2, are mounted on the front and rear of the roof, respectively. Both cameras utilize Charge-Coupled Device (CCD) sensors operating in global shutter mode. This offers advantages over rolling shutter cameras, particularly at high vehicle speeds. Global shutter mode exposes all pixels in the captured

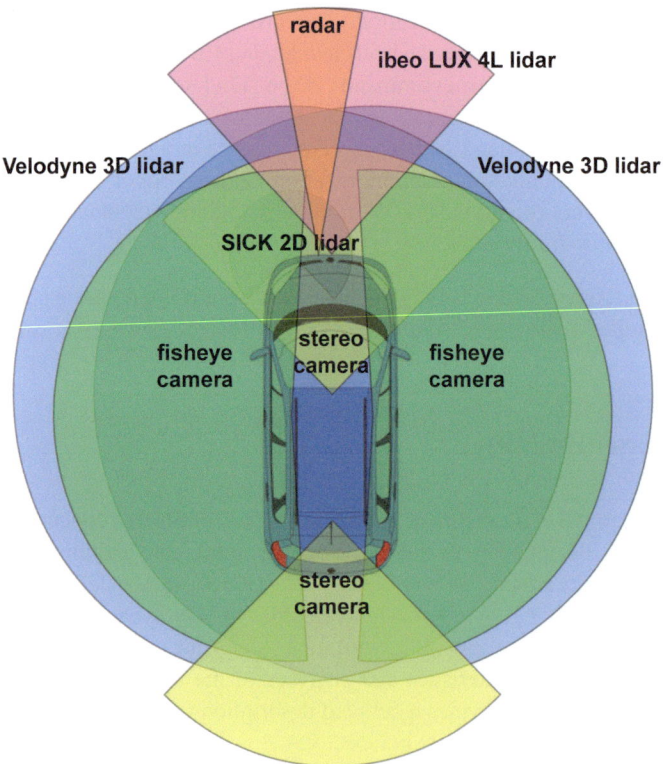

Fig. 2.4 Schematic diagram of the sensing areas of various sensors of the UTBM RoboCar

2.4 Dataset Building

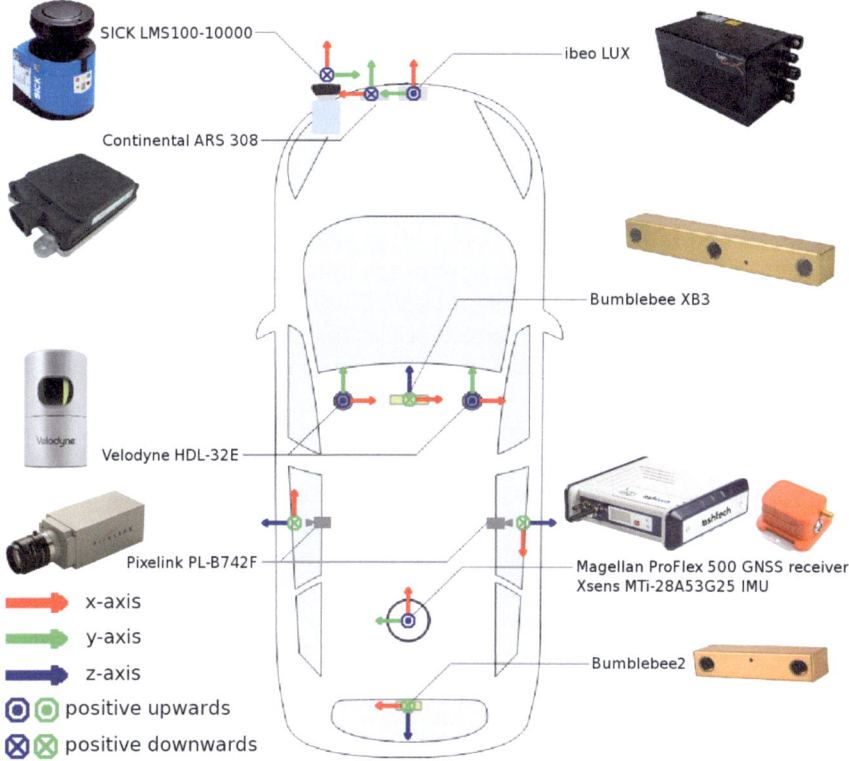

Fig. 2.5 The sensors used and their mounting positions

image simultaneously, whereas rolling shutter mode exposes pixels sequentially, typically in a wave-like pattern.

- *3D lidars*: Two Velodyne HDL-32E lidars are mounted side-by-side on the front of the roof. Each lidar features 32 scanning channels, a 360-degree horizontal and 40-degree vertical FoV, and a measurement range of up to 100 m. When multiple lidars are deployed in close proximity, as on the UTBM RoboCar, mutual interference can occur due to reflections. To mitigate this, the lidars' built-in phase-locking capabilities are employed to manage laser emission overlap, and post-processing techniques are applied to eliminate data shadows.
- *Fisheye cameras*: Two Pixelink PL-B742F industrial cameras, each equipped with a Fujinon FE185C086HA-1 fisheye lens, are mounted on either side of the mid-roof, facing laterally. These cameras utilize Complementary Metal-Oxide-Semiconductor (CMOS) global shutter sensors to capture high-speed motion without distortion. The fisheye lenses provide an ultra-wide 185-degree FoV. This configuration enhances lateral environmental perception and provides complementary color and texture information to the Velodyne lidars.

- *2.5D Lidar*: An ibeo LUX 4L lidar is integrated into the front bumper near the vehicle's y-axis, providing an 85-degree (four-layer mode) or 110-degree (two-layer mode) horizontal FoV and a measurement range of up to 200 m. This lidar is paired with a radar sensor, forming a crucial safety system for the vehicle and other road users.
- *Radar*: A Continental ARS 308 radar is mounted adjacent to the ibeo lidar. While radar offers lower angular accuracy compared to lidar, it provides robust detection of moving objects due to the Doppler effect and is less susceptible to adverse weather conditions. Some radar systems can even detect objects behind obstacles using reflections [37]. Current research is increasingly focused on 4D radar, which provides object height information in addition to traditional radar measurements [38]. On the UTBM RoboCar, radar and lidar data are cross-referenced for enhanced object detection and tracking.
- *Laser Rangefinder*: A SICK LMS100-10000 laser rangefinder is mounted on one side of the front bumper. This 2D lidar provides a 270-degree FoV and, due to its slight downward tilt, acquires data about the road surface, including markings and boundaries. The combined use of the ibeo and the SICK lidars is also recommended by the industry, with the former for object detection (i.e. dynamics) and the latter for road understanding (i.e. statics).
- *GNSS Receiver*: A Magellan ProFlex 500 GNSS receiver placed inside the vehicle, with two antennas mounted on the roof, is used for localization. One antenna is mounted along the vehicle's z-axis, perpendicular to the rear axle, for satellite signal reception, while the other is positioned at the rear of the roof for synchronization with a Real-Time Kinematic (RTK) base station. RTK correction enhances GNSS positioning accuracy from meter-level to centimeter-level.
- *IMU*: An Xsens MTi-28A53G25 Inertial Measurement Unit (IMU) is installed inside the vehicle to measure linear acceleration, angular velocity, and absolute orientation. IMUs are often used in conjunction with GNSS receivers.

For sensor data acquisition and processing, the ibeo lidar and radar, critical for driving safety, are connected to a dedicated control unit. This unit interfaces with the vehicle's Controller Area Network (CAN) bus to enable real-time vehicle control, including steering, acceleration, and braking. The two Velodyne lidars (via Ethernet) and the GNSS/IMU (via USB 2.0) are connected to an embedded computing unit, which performs core autonomous driving functions such as SLAM, point cloud clustering, sensor fusion, vehicle localization, and path planning. All cameras are connected via IEEE 1394 to a gaming laptop, providing the necessary computational resources, particularly GPU capacity, for vision algorithms. All sensors are connected via a wired network to a Dell Precision Tower 3620 workstation, which serves as the data acquisition and logging platform for dataset generation. To power the embodied perception platform for approximately one hour, two 60 Ah batteries were added, given the vehicle's diesel engine.

2.4 Dataset Building

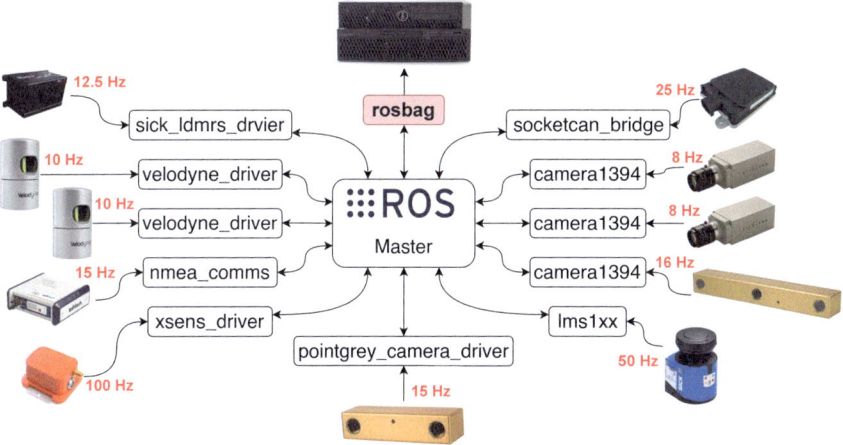

Fig. 2.6 Architecture diagram of the ROS-based software for data collection. To facilitate reproducibility, the diagram depicts the ROS package name for each sensor driver rather than the specific ROS node names. In practice, the ROS master communicates with the nodes provided by these packages

2.4.2 Software Architecture

The software that data collection relies on is entirely based on ROS. Its architecture and the data publication frequency of each sensor are illustrated in Fig. 2.6. All ROS nodes run locally on the Dell workstation to ensure software-level data synchronization via ROS timestamps. No data delays were observed during the acquisition process, primarily because the system records only raw data, deferring post-processing to offline playback. Data is stored in the "rosbag" format.

2.4.3 Sensor Calibration

All cameras and lidars were intrinsically calibrated, and the corresponding calibration files are included with the dataset. Camera calibration was performed using the ROS "camera_calibration" package with a chessboard pattern. The lidars were calibrated using factory-provided intrinsic parameters. Extrinsic lidar calibration was performed by minimizing the voxel-wise L_2 distance between point clouds from different sensors, acquired by driving the vehicle in a structured environment with multiple landmarks. To calibrate the transformation between the stereo cameras and the 3D lidars, the vehicle was positioned to face building corners, and the resulting point clouds were manually aligned on three planar surfaces (i.e., two walls and the ground). The aligned sensor data is shown in Fig. 2.7, demonstrating good alignment between the lidar and stereo camera point clouds.

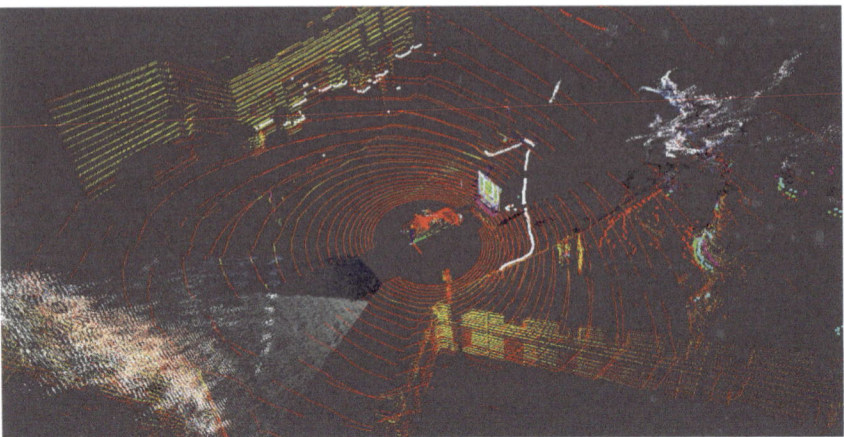

Fig. 2.7 A ROS Rviz screenshot of the data collected with calibrated sensors. The UTBM RoboCar is in the centre of the image with a truck in front. The red ring points come from the Velodyne lidars, the white points from the SICK lidar, and the colored points from the ibeo lidar. The point clouds in front of and behind the vehicle are from the two Bumblebe stereo cameras

2.4.4 Comparison Between Different Datasets

To contextualize the EU long-term dataset, a comparison with other contemporary open datasets is presented in Table 2.3. Notably, KITTI [2] is a pioneering dataset for autonomous driving research, providing stereoscopic color images, 3D lidar point clouds, and vehicle GPS coordinates. Two key factors contributed to its success. First, it offers extensive high-quality manual annotations, including categories such as "Car", "Van", "Truck", "Pedestrian", "Person (sitting)", "Cyclist", "Tram", and "Misc" (e.g., trailers, Segways), with each label further classified by difficulty: "easy", "moderate", or "hard". Second, it features an online benchmark ranking for various tasks, including stereo vision, optical flow, scene flow, visual odometry, object detection and tracking, road/lane detection, and semantic segmentation. Furthermore, several extensions of KITTI exist, such as KITTI-360 [39] and SemanticKITTI [33].

KITTI also presents certain limitations. For instance, the dataset employs one-to-one frame synchronization to handle sensor data with varying acquisition frequencies. Moreover, point cloud annotations are derived from image label projections, which may not accurately reflect the true geometric properties of the point cloud [44]. Additionally, KITTI predominantly features conventional scenes with low dynamic complexity and favorable weather and lighting conditions. These scenes do not fully represent the challenges encountered in real-life driving scenarios. Furthermore, the object detection and tracking benchmarks within the dataset exhibit temporal discontinuities and lack positioning information about the vehicle itself. In summary, while

2.4 Dataset Building

Table 2.3 A comparison of open datasets for autonomous driving

Dataset	Sensor	Synchronization	Ground truth	Location	Weather	Time
EU long-t [9]	2 × 32-layer lidar 1 × 4-layer lidar 1 × 1-layer lidar 2 × stereo camera 2 × fisheye camera 1 × radar 1 × GPS-RTK 1 × independent IMU	Software (ROS timestamp) and hardware (PPS for the two Velodynes)	GPS-RTK/IMU for vehicle self-localization	France[a]	Sun, clouds, snow	Day, dusk, night, three seasons (spring, summer, winter)
KITTI [2]	1 × 64-layer lidar 2 × grayscale camera 2 × color camera 1 × GPS-RTK/IMU	Software and hardware (reed contact)	Scene flow, odometry object detection and tracking, road and lane	Germany[a]	Clear	Day, autumn
Oxford [36]	1 × 4-layer lidar 2 × 1-layer lidar 1 × stereo camera 3 × fisheye camera 1 × GPS-RTK/INS	Software	GPS-RTK/INS for vehicle self-localization	UK[b]	Sun, clouds, overcast, rain snow	Day, dusk, night, four seasons
KAIST [40]	2 × 16-layer lidar 2 × 1-layer lidar 2 × monocular camera 1 × consumer-level GPS 1 × GPS-RTK 1 × fiber optics gyro 1 × independent IMU 2 × wheel encoder 1 × altimeter	Software (ROS timestamp) and hardware (PPS for the two Velodynes, an external trigger for the two monocular cameras to get stereo)	SLAM algorithm for vehicle self-localization	South Korea[a]	Clear	Day
ApolloScape [41]	2 × 1-layer lidar[c] 6 × monocular camera 1 × GPS-RTK/IMU	Unknown	Scene parsing, car instance, lane segmentation, self localization, detection and tracking, trajectory, stereo	China[a]	Unknown	Day
nuScenes [42]	1 × 32-layer lidar 6 × monocular camera 5 × radar 1 × GPS-RTK 1 × independent IMU	Software	HD map-based localization, object detection and tracking	US[a] Singapore[b]	Sun, clouds, rain	Day, night
Waymo [3]	5 × lidar[d] 5 × camera[d]	Unknown but very well-synchronized	Object detection and tracking	US[a]	Sun, rain	Day, night

(continued)

Table 2.3 (continued)

Dataset	Sensor	Synchronization	Ground truth	Location	Weather	Time
CADC [43]	1 × 32-layer lidar 8 × monocular camera 1 × GPS-RTK/IMU 2 × independent IMU 1 × ADAS kit	Hardware	Object detection and tracking	Canada[a]	Snowfall	Day

Dataset	Distance (km)	Data format	Baseline[e]	Download	License	Privacy	First release
EU long-t [9]	63.4	Rosbag (all-in-one)	2	Free	CC BY-NC-SA 4.0	Face and plate removed	Nov. 2018
KITTI [2]	39.2	Bin (lidar), png (camera) txt (GPS-RTK/IMU)	3	Registration	CC BY-NC-SA 3.0	Removal under request	Mar. 2012
Oxford [36]	1010.46	Bin (lidar), png (camera) csv (GPS-RTK/INS)	0	Registration	CC BY-NC-SA 4.0	Removal under request	Oct. 2016
KAIST [40]	190,989	Bin (lidar), png (camera) csv (GPS-RTK/IMU)	1	Registration	CC BY-NC-SA 4.0	Removal under request	Sep. 2017
ApolloScape [41]	Unknown	Png (lidar), jpg (camera)	1	Registration	ApolloScape License	Removal under request	Apr. 2018
nuScenes [42]	242	Xml	3	Registration	CC BY-NC-SA 4.0	Face and plate removed	Mar. 2019
Waymo [3]	Unknown	Range image (lidar) jpeg (camera)	3	Registration	Waymo License	Face and plate removed	Aug. 2019
CADC [43]	20	Bin (lidar), png (camera) txt (GPS-RTK/IMU/ADAS)	0	Registration	CC BY-NC 4.0	Removal under request	Jan. 2020

[a] Right-hand traffic,
[b] left-hand traffic,
[c] vertical scanning,
[d] device model undisclosed,
[e] only including methods published with the paper, excluding community contributions

KITTI is a valuable resource for computer vision-centric approaches to autonomous driving, it does not fully encompass the complexities of the autonomous driving task itself.

Another noteworthy dataset is Waymo [3], launched in 2019. Developed with significant industry resources, Waymo represents a substantial advancement in autonomous driving data. Organized by scene, each scene comprises continuous and annotated multi-modal sensor data, including real-time vehicle localization, facilitating a more comprehensive understanding of the driving environment. The continuous nature of the data is particularly beneficial for research in areas such as online learning, domain adaptation, object tracking, and trajectory prediction. To date, Waymo has released 798 scenes for model training and 202 scenes for model validation, each encompassing 20 seconds of continuous driving time. Data was collected in diverse urban and suburban locations, covering a range of lighting and weather conditions, including day, night, dawn, dusk, sunshine, and rain. Hardware-level synchronization ensures precise alignment between sensor data streams. Data annotation covers four object classes including cars, pedestrians, cyclists, and signs. However, the classification of motorcycles and motorcyclists as vehicles and scooter riders as pedestrians suggests a need for more granular classification in future releases.

2.5 Discussions

This chapter presented a study on benchmarking methodologies for mobile robotics. It addressed the crucial question of how to conduct benchmarking effectively by defining three key aspects of the process: parameters, metrics, and experimental design. Subsequently, two concrete examples—the development of a testbed and the construction of a dataset—illustrated potential benchmarking platforms. However, benchmarking has faced evolving challenges, four of which are discussed below.

2.5.1 Ranking-Driven Overfitting

As discussed in the introduction, benchmarking, particularly those benchmarks with public rankings, can be a double-edged sword. While offering numerous benefits, they can also induce overfitting to a specific dataset or evaluation metric. A practical consequence is that methods reporting superior performance on one benchmark suite may not generalize well to others. For example, point cloud annotations in the KITTI dataset (i.e., 3D bounding boxes of various road users) are derived from 2D image projections. Annotators manually label images, and the corresponding point clouds are then automatically annotated using a calibration between the two modalities. Conversely, point cloud annotation in the L-CAS 3D Point Cloud People Dataset is performed directly on the point clouds. This difference in annotation methodology results in KITTI's 3D bounding boxes including object size estimates, while L-CAS bounding boxes do not. Consequently, when evaluating point cloud clustering or object detection methods on KITTI, object size estimation can improve performance. However, when evaluating on L-CAS, this additional step can have a detrimental effect. This is just one example of potential overfitting, a more common approach is to fine-tune models based on dataset characteristics.

2.5.2 Benchmarking AI with AI

Using AI to evaluate AI is a long-standing concept, echoing the Turing test's principle of using inter-agent interaction for evaluation. As an evaluator, a symbolic approach offers logical power but struggles to establish a one-to-one mapping of object properties and effectively handle underlying representations. With the advancement of modern AI, particularly Deep Neural Network (DNN), connectionism has achieved significant breakthroughs, especially in addressing problems that challenged symbolic methods. However, a recognized limitation of connectionism is its lack of interpretability. Symbolism and connectionism can be analogously compared to human rational and perceptual thinking, respectively. The former emphasizes analysis and reasoning, while the latter involves emotion and intuition. A promising research direction lies therefore in combining these two paradigms to develop more robust and capable AI systems.

This chapter's content incorporated the concept of benchmarking AI with AI. The testbed described in Sect. 2.3, grounded in symbolic reasoning, facilitated comparative analysis by addressing questions such as "In what aspects is method A superior to method B?" or "Under what conditions is method A (or method B) more appropriate?" The work presented in Sect. 2.2.2.3 further explored this concept by investigating the correlation between RCM and HCM to evaluate the social acceptability of robot navigation—indirectly considering human feelings—using only RCM when HCM is unavailable. This represented an example of using AI to evaluate HRI, employing a statistical approach. Future work could involve developing dependency functions using neural networks, although this would necessitate the acquisition of additional training data.

2.5.3 Benchmarking Ethics

In principle, benchmarking should adhere to established ethical principles, including the avoidance of bias, promotion of transparency, and minimization of harm. The research practices relevant to this book primarily involve experiments with human participants. Considering socially-compliant robot navigation as an example, such experiments must adhere to the following ethical considerations:

- *Safety*: Benchmarking must prioritize the safety of human participants. This entails requirements for the mobile robot's obstacle detection and avoidance capabilities, as well as rapid response times in emergency situations. Furthermore, experimental designs should incorporate protective measures for participant safety, such as emergency braking systems.
- *Fairness*: Benchmarking should ensure equitable treatment of all participants, regardless of race, gender, socioeconomic status, or other protected attributes. This involves verifying that the algorithms deployed on the robots do not exhibit bias in their decision-making processes and ensuring fair participant recruitment and experimental design practices.
- *Transparency*: The benchmarking process should be transparent and readily understandable to participants. This requires providing participants with detailed explanations of the experimental procedures prior to commencement, including information regarding the collection, storage, and use of personal information and experimental data, as well as explanations of the robot's behavior and safety protocols for participant self-protection.
- *Accountability*: Clear lines of responsibility for the conduct of benchmarking activities must be established.

2.5.4 Data Privacy

Mobile robotics, particularly in domains such as autonomous vehicles and delivery drones, often entails the collection of substantial amounts of data. This data may include sensor readings, location information, and potentially personally identifiable information (PII) if the robots interact with people or their environments. Data privacy is therefore a critical concern in benchmarking, aiming to protect sensitive information while preserving the integrity of the benchmarking process. Reconciling these two objectives, however, remains challenging. For instance, in the EU long-term dataset [9], while anonymization or pseudonymization techniques can enhance privacy, they may also introduce noise or distortions that compromise the accuracy and reliability of benchmarking results. Furthermore, obtaining informed consent from all individuals potentially affected by benchmarking in public settings is often impractical and resource-intensive [13]. Moreover, even with robust security measures in place, the risk of data breaches or unauthorized access persists, potentially compromising both privacy and benchmarking integrity. Balancing the pursuit of accurate benchmarking with the need to safeguard individual privacy presents ethical dilemmas, especially when dealing with sensitive data. Consequently, recommended practices include incorporating privacy considerations into the benchmarking process from the outset, exploring advanced privacy-preserving techniques such as differential privacy or federated learning to minimize data exposure, and consulting ethics review boards to ensure adherence to ethical principles in all benchmarking activities [19].

References

1. Burgard, W., Stachniss, C., Grisetti, G., Steder, B., Kümmerle, R., Dornhege, C., Ruhnke, M., Kleiner, A., Tardós, J.D.: A comparison of SLAM algorithms based on a graph of relations. In: IEEE/RSJ International Conference on Intelligent Robots and Systems (IROS), pp. 2089–2095 (2009). https://doi.org/10.1109/IROS.2009.5354691
2. Geiger, A., Lenz, P., Urtasun, R.: Are we ready for autonomous driving? the KITTI vision benchmark suite. In: IEEE/CVF Conference on Computer Vision and Pattern Recognition (CVPR), pp. 3354–3361 (2012). https://doi.org/10.1109/CVPR.2012.6248074
3. Sun, P., Kretzschmar, H., Dotiwalla, X., Chouard, A., Patnaik, V., Tsui, P., Guo, J., Zhou, Y., Chai, Y., Caine, B., Vasudevan, V., Han, W., Ngiam, J., Zhao, H., Timofeev, A., Ettinger, S., Krivokon, M., Gao, A., Joshi, A., Zhang, Y., Shlens, J., Chen, Z., Anguelov, D.: Scalability in perception for autonomous driving: Waymo open dataset. In: IEEE/CVF Conference on Computer Vision and Pattern Recognition (CVPR), pp. 2443–2451 (2020). https://doi.org/10.1109/CVPR42600.2020.00252
4. Yan, Z., Fabresse, L., Laval, J., Bouraqadi, N.: Building a ros-based testbed for realistic multi-robot simulation: taking the exploration as an example. Robotics **6**(3), 1–21 (2017)
5. Yan, Z., Fabresse, L., Laval, J., Bouraqadi, N.: Team size optimization for multi-robot exploration. In: Proceedings of the 4th International Conference on Simulation, Modeling, and Programming for Autonomous Robots (SIMPAR), pp. 438–449. Bergamo, Italy (2014)
6. Yan, Z., Jouandeau, N., Ali Cherif, A.: A survey and analysis of multi-robot coordination. Int. J. Adv. Rob. Syst. **10**(399), 1–18 (2013)
7. Yang, R., Yan, Z., Yang, T., Ruichek, Y.: Efficient online transfer learning for 3d object classification in autonomous driving. In: Proceedings of the 2021 IEEE International Conference on Intelligent Transportation Systems (ITSC), pp. 2950–2957. Indianapolis, USA (2021)
8. Yang, R., Yan, Z., Yang, T., Wang, Y., Ruichek, Y.: Efficient online transfer learning for road participants detection in autonomous driving. IEEE Sens. J. **23**(19), 23522–23535 (2023)
9. Yan, Z., Sun, L., Krajnik, T., Ruichek, Y.: EU long-term dataset with multiple sensors for autonomous driving. In: Proceedings of the 2020 IEEE/RSJ International Conference on Intelligent Robots and Systems (IROS), pp. 10697–10704. Las Vegas, USA (2020)
10. Yan, Z., Duckett, T., Bellotto, N.: Online learning for human classification in 3D LiDAR-based tracking. In: Proceedings of the 2017 IEEE/RSJ International Conference on Intelligent Robots and Systems (IROS), pp. 864–871. Vancouver, Canada (2017)
11. Majer, F., Yan, Z., Broughton, G., Ruichek, Y., Krajnik, T.: Learning to see through haze: radar-based human detection for adverse weather conditions. In: Proceedings of the 2019 European Conference on Mobile Robots (ECMR), pp. 1–7. Prague, Czech Republic (2019)
12. Cosar, S., Yan, Z., Lambrou, T., Yue, S., Bellotto, N.: Thermal camera based physiological monitoring with an assistive robot. In: Proceedings of the 40th Annual International Conference of the IEEE Engineering in Medicine and Biology Society (EMBC), pp. 1–4. Honolulu, USA (2018)
13. Okunevich, I., Hilaire, V., Galland, S., Lamotte, O., Shilova, L., Ruichek, Y., Yan, Z.: Human-centered benchmarking for socially-compliant robot navigation. In: Proceedings of the 2023 European Conference on Mobile Robots (ECMR), pp. 1–7. Coimbra, Portugal (2023)
14. Burgard, W., Moors, M., Fox, D., Simmons, R.G., Thrun, S.: Collaborative multi-robot exploration. In: IEEE International Conference on Robotics and Automation (ICRA), pp. 476–481 (2000). https://doi.org/10.1109/ROBOT.2000.844100
15. Stachniss, C.: Robotic Mapping and Exploration, Springer Tracts in Advanced Robotics, vol. 55. Springer, New York (2009). https://doi.org/10.1007/978-3-642-01097-2
16. Yan, Z., Jouandeau, N., Ali Cherif, A.: Multi-robot heuristic goods transportation. In: Proceedings of the 6th IEEE International Conference on Intelligent Systems (IS), pp. 409–414. Sofia, Bulgaria (2012)
17. Zlot, R., Stentz, A., Dias, M.B., Thayer, S.: Multi-robot exploration controlled by a market economy. In: IEEE International Conference on Robotics and Automation (ICRA), pp. 3016–3023 (2002). https://doi.org/10.1109/ROBOT.2002.1013690

References

18. Padilla, R., Netto, S.L., da Silva, E.A.B.: A survey on performance metrics for object-detection algorithms. In: International Conference on Systems, Signals and Image Processing (IWSSIP), pp. 237–242 (2020). https://doi.org/10.1109/IWSSIP48289.2020.9145130
19. Okunevich, I., Lombard, A., Krajnik, T., Ruichek, Y., Yan, Z.: Online context learning for socially compliant navigation. IEEE Robot. Autom. Lett. **14**, 1–8 (2025)
20. Chen, Y.F., Everett, M., Liu, M., How, J.P.: Socially aware motion planning with deep reinforcement learning. In: IEEE/RSJ International Conference on Intelligent Robots and Systems (IROS), pp. 1343–1350 (2017). https://doi.org/10.1109/IROS.2017.8202312
21. Everett, M., Chen, Y.F., How, J.P.: Motion planning among dynamic, decision-making agents with deep reinforcement learning. In: IEEE/RSJ International Conference on Intelligent Robots and Systems (IROS), pp. 3052–3059 (2018). https://doi.org/10.1109/IROS.2018.8593871
22. Liu, L., Dugas, D., Cesari, G., Siegwart, R., Dubé, R.: Robot navigation in crowded environments using deep reinforcement learning. In: IEEE/RSJ International Conference on Intelligent Robots and Systems (IROS), pp. 5671–5677 (2020). https://doi.org/10.1109/IROS45743.2020.9341540
23. Gao, Y., Huang, C.: Evaluation of socially-aware robot navigation. Front. Robot. AI **8**, 721317 (2021). https://doi.org/10.3389/FROBT.2021.721317
24. Mavrogiannis, C.I., Hutchinson, A.M., Macdonald, J., Alves-Oliveira, P., Knepper, R.A.: Effects of distinct robot navigation strategies on human behavior in a crowded environment. In: Proceedings of the 14th ACM/IEEE International Conference on Human-Robot Interaction (HRI), pp. 421–430 (2019). https://doi.org/10.1109/HRI.2019.8673115
25. Lu, D.V., Smart, W.D.: Towards more efficient navigation for robots and humans. In: IEEE/RSJ International Conference on Intelligent Robots and Systems (IROS), pp. 1707–1713 (2013). https://doi.org/10.1109/IROS.2013.6696579
26. Carpinella, C.M., Wyman, A.B., Perez, M.A., Stroessner, S.J.: The robotic social attributes scale (rosas): development and validation. In: ACM/IEEE International Conference on Human-Robot Interaction (HRI), pp. 254–262 (2017). https://doi.org/10.1145/2909824.3020208
27. Pan, M.K.X.J., Croft, E.A., Niemeyer, G.: Evaluating social perception of human-to-robot handovers using the robot social attributes scale (rosas). In: ACM/IEEE International Conference on Human–Robot Interaction (HRI), pp. 443–451 (2018). https://doi.org/10.1145/3171221.3171257
28. Vintr, T., Blaha, J., Rektoris, M., Ulrich, J., Roucek, T., Broughton, G., Yan, Z., Krajnik, T.: Toward benchmarking of long-term spatio-temporal maps of pedestrian flows for human-aware navigation. Front. Robot. AI **9**, 890013 (2022)
29. Yan, Z., Fabresse, L., Laval, J., Bouraqadi, N.: Benchmarking de performance pour l'exploration multi-robots. In: Proceedings of the Vingt-troisièmes Journées Francophones sur les Systèmes Multi-Agents (JFSMA), pp. 9–18. Rennes, France (2015)
30. Yan, Z., Fabresse, L., Laval, J., Bouraqadi, N.: Metrics for performance benchmarking of multi-robot exploration. In: Proceedings of the 2015 IEEE/RSJ International Conference on Intelligent Robots and Systems (IROS), pp. 3407–3414. Hamburg, Germany (2015)
31. Yan, Z., Fabresse, L., Laval, J., Bouraqadi, N.: Benchmarking des performances de systèmes multirobots: application à l'exploration. Revue d'Intelligence Artificielle **30**(1–2), 211–236 (2016)
32. Echeverria, G., Lemaignan, S., Degroote, A., Lacroix, S., Karg, M., Koch, P., Lesire, C., Stinckwich, S.: Simulating complex robotic scenarios with MORSE. In: Proceedings of the 3rd International Conference on Simulation, Modeling, and Programming for Autonomous Robots (SIMPAR), pp. 197–208 (2012). https://doi.org/10.1007/978-3-642-34327-8_20
33. Behley, J., Garbade, M., Milioto, A., Quenzel, J., Behnke, S., Gall, J., Stachniss, C.: Towards 3d lidar-based semantic scene understanding of 3d point cloud sequences: the semantickitti dataset. Int. J. Robot. Res. **40**(8–9), 735 (2021). https://doi.org/10.1177/02783649211006735
34. Santos, J.M., Portugal, D., Rocha, R.P.: An evaluation of 2d SLAM techniques available in robot operating system. In: IEEE International Symposium on Safety, Security, and Rescue Robotics (SSRR), pp. 1–6 (2013). https://doi.org/10.1109/SSRR.2013.6719348

35. Quigley, M., Conley, K., Gerkey, B.P., Faust, J., Foote, T., Leibs, J., Wheeler, R., Ng, A.Y.: ROS: an open-source robot operating system. In: ICRA Workshop on Open Source Software (2009)
36. Maddern, W., Pascoe, G., Linegar, C., Newman, P.: 1 year, 1000 km: the oxford robotcar dataset. Int. J. Robot. Res. **36**(1), 3–15 (2017). https://doi.org/10.1177/0278364916679498
37. Li, J., Zeng, Z., Sun, J., Liu, F.: Through-wall detection of human being's movement by UWB radar. IEEE Geosci. Remote Sens. Lett. **9**(6), 1079–1083 (2012). https://doi.org/10.1109/LGRS.2012.2190707
38. Fan, L., Wang, J., Chang, Y., Li, Y., Wang, Y., Cao, D.: 4d mmwave radar for autonomous driving perception: A comprehensive survey. IEEE Trans. Intell. Vehic. **9**(4), 4606–4620 (2024). https://doi.org/10.1109/TIV.2024.3380244
39. Liao, Y., Xie, J., Geiger, A.: KITTI-360: a novel dataset and benchmarks for urban scene understanding in 2d and 3d. IEEE Trans. Pattern Anal. Mach. Intell. **45**(3), 3292–3310 (2023). https://doi.org/10.1109/TPAMI.2022.3179507
40. Jeong, J., Cho, Y., Shin, Y., Roh, H.C., Kim, A.: Complex urban dataset with multi-level sensors from highly diverse urban environments. Int. J. Robot. Res. **38**(6), 996 (2019). https://doi.org/10.1177/0278364919843996
41. Huang, X., Cheng, X., Geng, Q., Cao, B., Zhou, D., Wang, P., Lin, Y., Yang, R.: The apolloscape dataset for autonomous driving. In: IEEE Conference on Computer Vision and Pattern Recognition Workshops (CVPR Workshops), pp. 954–960 (2018). https://doi.org/10.1109/CVPRW.2018.00141
42. Caesar, H., Bankiti, V., Lang, A.H., Vora, S., Liong, V.E., Xu, Q., Krishnan, A., Pan, Y., Baldan, G., Beijbom, O.: nuscenes: A multimodal dataset for autonomous driving. In: IEEE/CVF Conference on Computer Vision and Pattern Recognition (CVPR), pp. 11618–11628 (2020). https://doi.org/10.1109/CVPR42600.2020.01164
43. Pitropov, M., Garcia, D.E., Rebello, J., Smart, M., Wang, C., Czarnecki, K., Waslander, S.L.: Canadian adverse driving conditions dataset. Int. J. Robot. Res. **40**(4–5), 368 (2021). https://doi.org/10.1177/0278364920979368
44. Yang, T., Li, Y., Zhao, C., Yao, D., Chen, G., Sun, L., Krajnik, T., Yan, Z.: 3D ToF LiDAR in mobile robotics: a review. CoRR abs/2202.11025 (2022). URL http://arxiv.org/abs/2202.11025

Chapter 3
Robot Perception

Abstract This chapter presents a study on robot perception. It begins by establishing the research motivation: enabling large-scale human detection and tracking in public (non-domestic) environments using embodied sensors and onboard computing. Subsequently, it introduces contemporary 3D lidar technology as an embodied sensor, covering its fundamental operating principles and relevant applications in mobile robotics. The chapter then details the "adaptive clustering" method developed by the author, highlighting its advantages and limitations through a performance comparison with other established methods. Following this, it describes several hand-crafted features extracted from point clouds, proven effective for human model training. Finally, it presents a multi-target tracker optimized for point cloud data.

Keywords 3D lidar · Point clouds · Adaptive clustering · Object detection · Multi-target tracking

3.1 Introduction

Robot perception aims to equip robots with the ability to perceive the external world and their own state, analogous to human perception. This capability relies on a variety of sensors. The author's research has focused on the use of non-visual (active) sensors, including sonar, lidar, and radar, driven by four key motivations. First, distance information of objects within the environment is crucial for robot navigation. Second, effective perception is essential for robots operating in large-scale, highly dynamic environments. Third, long-term robot autonomy necessitates robust perception. Finally, robots deployed outdoors must be able to operate reliably in diverse weather conditions. This chapter presents work on embodied perception using 3D lidar. This research began in late 2015 within the framework of the European FLOBOT project, which involved integrating a 3D lidar onto an indoor cleaning robot, innovatively employing it for both human perception and robot navigation.

3.2 3D Lidar

Lidar, an acronym for "Light Detection And Ranging", is a technology that uses pulsed light, typically from a laser, to measure distances. Historically, research on how mobile robots perceive their external world has progressed through several stages, evolving from sonar and planar laser rangefinders to visual sensors and, more recently, 3D lidar and 4D millimeter-wave radar. Among them, the planar laser rangefinder played a crucial role in advancing Simultaneous Localization And Mapping (SLAM), one of the fontamental problems in mobile robotics, by providing accurate geometric representations of environments [53]. Over the past decade, 3D lidar, which shares the same ranging principle as planar laser rangefinders, has garnered increasing attention in both academia and industry. This trend is evidenced by the growing number of relevant publications (see Fig. 3.1), capital investment, and industry participants (see Fig. 3.2).

Compared with planar laser rangefinders, 3D lidar not only provides an additional spatial dimension of information by increasing the number of scanning layers, but also enables wider and longer distance measurements. For instance, the commercially available Robosense Ruby Plus offers a 360-degree horizontal and 40-degree vertical FoV, with a measurement range of up to 250 m. This enhanced sensing capability allows researchers to investigate more complex problems in larger and more intricate environments, such as large-scale human detection and tracking in public spaces like airports, supermarkets, and cafeterias (see Fig. 3.3) [56–58].

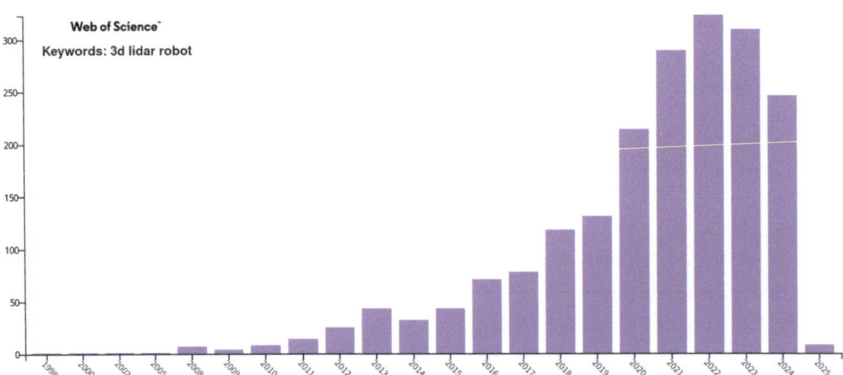

Fig. 3.1 Number of publications per year related to "3D lidar robot" indexed in Web of Science, as of 2024

3.2 3D Lidar

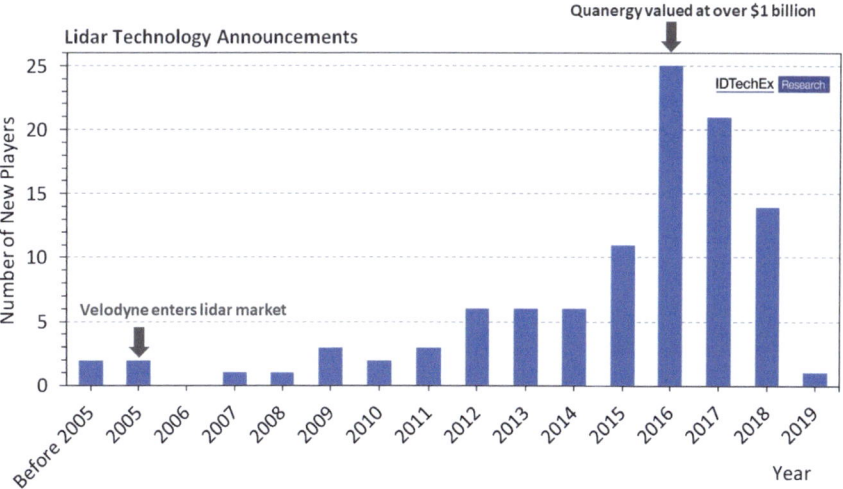

Fig. 3.2 Number of new 3D lidar companies entering the market each year until 2019, based on IDTechEx statistics

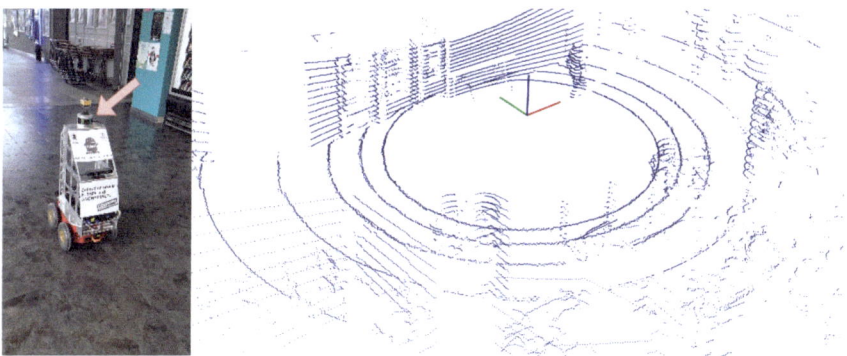

Fig. 3.3 Left: A robot equipped with a 3D lidar (indicated by the arrow) deployed in a university canteen for human detection and tracking. Right: Point cloud data generated by the 3D lidar

3.2.1 Ranging Principle

This book focuses on 3D lidar based on the Time-of-Flight (ToF) principle, illustrated in Fig. 3.4 [68]. Specifically, a laser transmitter emits pulsed laser light, typically with a wavelength between 905 and 1550 nm, in a specific direction. Upon encountering an obstacle, the laser beam is reflected or scattered, depending on the object's surface material. A laser detector then receives the returned signal and determines the distance between the sensor and the object by measuring the time of flight. This measurement mechanism is expressed as:

$$r = \frac{c \cdot \Delta t}{2n} \tag{3.1}$$

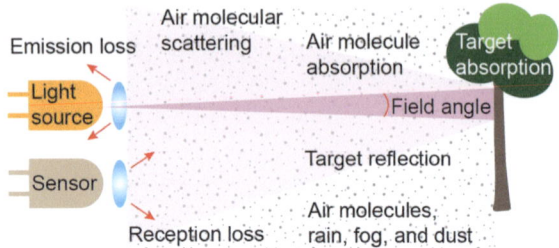

Fig. 3.4 Schematic of the ToF lidar ranging principle

where c represents the speed of light (a fundamental physical constant), n denotes the refractive index of the propagation medium, and Δt is the time difference between the transmission and reception of the laser pulse.

However, laser beam reception is not as straightforward as its emission. The received laser power is governed by the lidar equation, which can be expressed as [39]:

$$P_R = C \frac{\beta}{R^2} \exp\left(-2 \int_0^R \alpha(r) dr\right) \tag{3.2}$$

where: P_R is the received laser power at distance R; C is a constant incorporating factors such as the speed of light, laser transmit power, optical aperture area of the detector, and overall system efficiency; β is the target reflectivity (reflection efficiency of the object surface); α is the extinction coefficient of the lidar signal. As this equation demonstrates, sophisticated signal processing techniques are required to detect the true return signal in the presence of low signal-to-noise ratios.

3.2.2 Scanning Architectures

Commercially available 3D ToF lidars can be categorized into three types based on their scanning architectures (see Fig. 3.5): mechanical, semi-solid-state (also known as hybrid solid-state), and solid-state. Mechanical lidars employ multiple vertically arranged laser beams and a motor to rotate the entire optoelectronic assembly 360°C. Their advantages include mature technology and high measurement accuracy. However, they are characterized by larger size and higher cost. Semi-solid-state lidars feature a fixed transceiver module and a mechanically moving scanner. Two mainstream technical solutions exist including rotating mirror-based and Micro-Electro-Mechanical Systems (MEMS)-based. Compared to mechanical lidars, semi-solid-state lidars have fewer moving parts, resulting in greater stability and lower manufacturing costs, but typically offer a narrower FoV. Compared to solid-state lidars,

3.2 3D Lidar

Fig. 3.5 Three different types of ToF lidar: mechanical (left), semi-solid-state (middle), and solid-state (right)

semi-solid-state technology is relatively mature and thus more readily commercialized. Solid-state lidars contain no mechanical moving components. Current mainstream technical solutions include Flash lidar and Optical Phased Array (OPA)-based lidar. Solid-state lidars offer advantages such as smaller size and lower cost, but currently exhibit limitations in measurement accuracy and FoV, and require further technological maturation.

For prevalent mechanical lidars, increasing the number of scanning layers, and thus vertical resolution, necessitates stacking laser transceiver modules, leading to increased cost and manufacturing complexity. Consequently, while mechanical lidars remain widely adopted due to their high technological maturity and 360-degree horizontal FoV, industry and academia are increasingly investing in MEMS lidar, particularly for autonomous driving applications. However, the small reflector and receiver aperture of MEMS lidar limit its detection range, and further iterations of both the transceiver and scanning modules are required. Flash lidar illuminates an area with a single, diffused laser pulse. Due to human eye safety regulations on laser emission power, flash lidar struggles to achieve long-range measurements and is therefore primarily employed in mid-range or indoor applications.

In addition to the aforementioned ToF-based lidars, Frequency Modulated Continuous Wave (FMCW) lidar represents a promising alternative. Unlike ToF lidar, FMCW lidar transmits and receives continuous laser beams, mixes the returned light with a local oscillator signal, and employs heterodyne detection to measure the frequency difference between the transmitted and received signals. This frequency difference is then used to calculate the target distance. In principle, FMCW-based measurements offer greater stability and reliability compared to ToF-based measurements. However, FMCW technology is still undergoing active development and refinement. For a detailed comparison of various lidar technologies, the reader is referred to [25].

3.2.3 Physical Properties

A key advantage of lidar is its rapid data acquisition and high-precision range measurement capabilities. Among various types, mechanical lidar, commonly employed

Table 3.1 A comparison of the performance of commonly used exteroceptive sensors

Sensor	Lidar	Radar	Sonar	Camera
Detection range	★★★	★★★	★	★★
Ranging accuracy	★★★	★★	★★	★
Resolution	★★	★	★	★★★
Horizontal FoV	★★★	★★	★	★★
Vertical FoV	★★	★	★	★★
Color information	★★	★	★	★★★
Lighting robustness	★★★	★★★	★★★	★
Weather resistance	★★	★★★	★★★	★

in mobile robotics, offers 360-degree horizontal scanning and measurement ranges extending to hundreds of meters. Furthermore, as an active sensor, lidar exhibits robustness to varying lighting conditions, contributing to long-term robot autonomy. However, lidar data is typically represented as sparse point clouds that become sparser with increasing distance, and lacks readily interpretable features such as color and texture, posing challenges for object recognition. From another perspective, this inherent difficulty in object and human identification can be viewed as a positive attribute for privacy protection. Another limitation of lidar is its susceptibility to adverse weather conditions, primarily due to atmospheric water droplets. These droplets both absorb and scatter near-infrared laser light, increasing the extinction coefficient α (cf. Eq. 3.2). Additionally, wet surfaces reduce object reflectivity β [22]. The combined effect of these factors diminishes the received laser power, hindering object detection. Moreover, raindrops and snowflakes near the laser transmitter introduce significant measurement noise [8]. A performance comparison of various exteroceptive sensors, to better understand the physical properties of lidar, is provided in Table 3.1.

3.2.4 Data Representation

Lidar's measurements of its surroundings can be represented by a set of points in a three-dimensional coordinate space:

$$P = \{p_i \mid p_i = (x_i, y_i, z_i) \in \mathbb{R}^3, i = 1, \ldots, I\} \quad (3.3)$$

where each point p_i corresponds to a processed laser beam reflection, and I denotes the total number of points. By convention, this set of points is called a *point cloud*.

Depending on the characteristics of the hardware, additional attributes such as intensity and laser ring number may be associated with each point.

A widely used point cloud processing library in robotics is the Point Cloud Library (PCL) [43]. This open-source C++ library leverages traditional computer vision algorithms (excluding Deep Neural Network (DNN)) for tasks like feature estimation, surface reconstruction, 3D registration, model fitting, and segmentation. Notably, the Point Cloud Data (PCD) file format[1] serves as the native file format for PCL, facilitating efficient storage and exchange of point cloud data.

Another method for representing and processing lidar data leverages the Robot Operating System (ROS). ROS has emerged as the *de facto* standard platform for robot software development, and its influence extends to the realm of autonomous driving software, with a growing number of researchers and companies adopting it as the foundation. Within the ROS framework, lidar data are commonly represented using the PointCloud2 data structure.[2] This structure facilitates the storage and sharing of lidar data through packaging into "rosbags" [56, 61]. Notably, the close relationship between ROS and PCL enables seamless handover of point cloud data processing tasks to PCL. Beyond these two prominent representations, other formats are also commonly used, including binary files (e.g., .bin) employed in datasets like KITTI [15], Oxford RobotCar [31], KAIST [17], and CADC [37], and XML files exemplified by nuScenes [9].

3.2.5 Industrial Applications

The widespread adoption of 3D lidar can be traced back to the success of Stanford University's Junior robot car [32] in the 2007 DARPA Grand Challenge, where a Velodyne's 64-layer 3D lidar served as the primary sensor for obstacle detection, including pedestrians, signposts, and other vehicles. Junior's success directly influenced and accelerated the development and application of 3D lidar in industry, particularly within the autonomous driving sector [9, 15, 17, 24, 25, 31, 37, 51, 61, 63, 66, 67]. Despite Tesla's reluctance to acknowledge its widespread use, 3D lidar has become the *de facto* standard for self-driving vehicles [1]. Beyond obstacle detection, 3D lidar is also employed for creating high-definition maps and performing localization based on these maps. These high-resolution maps incorporate detailed road information, such as lanes, traffic signs, and traffic rules. To achieve more comprehensive environmental measurements and minimize blind spots, autonomous vehicles often utilize multiple lidars [17, 49, 51, 61]. However, just as relying solely on vision-based solutions is insufficient, autonomous driving based solely on 3D lidar presents limitations. Consequently, multi-modal perception remains the dominant approach (cf. Sect. 2.4).

---

[1] http://pointclouds.org/documentation/tutorials/pcd_file_format.php.

[2] http://docs.ros.org/melodic/api/sensor_msgs/html/msg/PointCloud2.html.

 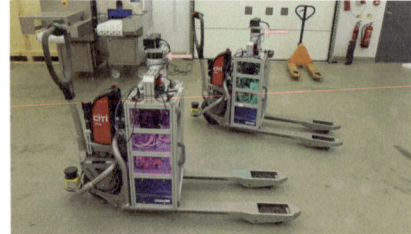

Fig. 3.6 Two types of service robots equipped with 3D lidar (pointed out by the red arrow)

Beyond autonomous driving, 3D lidar has found increasing application in service robotics for tasks such as professional cleaning [58], warehouse logistics [28], last-mile delivery [50], inspection [54], exploration [41], search and rescue [41], and agriculture [14]. Typically mounted on the robot's top to minimize occlusion, it facilitates tasks such as mapping, localization, object detection and tracking, semantic segmentation, and scene understanding. A concrete example is shown on the left side of Fig. 3.6. Within the EU-funded FLOBOT project, a robotic scrubber-dryer was developed for use in public spaces such as supermarkets, airports, and hospitals [58]. This robot employs a multi-modal embodied perception system, in which a 3D lidar is used for large-scale long-range human detection and tracking to ensure human safety when the robot is operating, and is also used for learning the changing patterns of the environment to optimize cleaning task planning. Another concrete example, depicted on the right side of Fig. 3.6, is the robotic forklift developed for warehouse logistics within the EU-funded ILIAD project [28]. This robot also utilizes a multi-modal embodied perception system, incorporating a 3D lidar for online mapping while eliminating human detection in the environment. A 2D obstacle grid map is then extracted from the 3D map for robot navigation and motion planning.

Furthermore, in the face of the last global public health crisis COVID-19, 3D lidar demonstrated encouraging results and broad application prospects. Initially, its most direct application involved human detection and tracking, enabling effective monitoring of social distancing while respecting individual privacy [45]. Subsequent research demonstrated the capability to detect mask usage [30]. Moreover, when integrated with other quarantine equipments, such as thermal sensors, 3D lidar can facilitate the detection and tracking of infected individuals and their contacts, enabling rapid and timely deployment of appropriate epidemic prevention measures. Last but not least, the use of 3D lidar in these applications is flexible: it can be mounted on mobile robots or deployed as a stationary sensor within the environment.

3.3 Object Detection in Point Clouds

So far, the main methods of object detection based on 3D lidar can be roughly divided into two categories including pipeline and end-to-end. The former means that different functional blocks (often termed modules) are interconnected into a pipeline. Therefore, the point cloud is first input into the pipeline, then processed by each module in sequence, and finally the pipeline outputs the detection results. For instance, the method to be introduced in this section involves initially segmenting the point cloud into non-overlapping subsets, ideally with each subset representing a distinct object. Subsequently, each subset is assigned a category label based on a specific model. This model can be either top-down, such as those based on machine learning [20, 56, 57, 59], or bottom-up, such as those based on object motion [12, 47].

End-to-end, on the other hand, is a modern approach closely related to deep learning methods that allows models to recognize objects directly from point clouds [21, 73]. Although this type of method has achieved results that break through the performance bottleneck of traditional ones in certain detection tasks, the current lack of model interpretability and the inability to domain shift make pipeline-based methods still irreplaceable. As evidence, rule-based clustering methods capable of detecting objectless objects are still widely used in the field of mobile robotics [7, 18, 43, 56, 71]. This section focuses on pipeline-based object detection, but first presents some representative works on end-to-end methods in the context of mobile robotics.

Currently, deploying end-to-end DNN-based methods onto robots still requires consideration of the computing capabilities of edge devices to a certain extent. An effective way to reduce the amount of computation is to convert 3D point clouds into 2D data. For example, PIXOR [62] first converts the 3D point cloud into a Bird's Eye View (BEV), while the latter is a 2D planar representation encoding each point in the former based on two channels including height and intensity, and then uses the structurally fine-tuned RetinaNet [26] on the BEV for object detection. Another example is Complex-YOLO [48], which first converts a 3D point cloud into a 2D BEV with height, intensity and density as channels (similar to how RGB images are encoded), then use YOLO [40] on the latter for object detection.

Another approach to enhance computational efficiency is to voxelize the 3D point cloud. For example, VoxelNet [73] first divides the 3D point cloud into multiple voxels, then randomly samples and normalizes the contained points, then uses several Voxel Feature Encoding layers to extract local features for each non-empty voxel, then these features are further abstracted through 3D Convolutional Middle Layers, and finally a Region Proposal Network is used for object detection. The subsequent improvement, SECOND [55], replaces VoxelNet's standard 3D convolution with sparse 3D convolution, achieving further gains in detection speed and memory efficiency. In addition, SWFormer [52] combines the two methods of BEV and voxelization, leveraging a Sparse Window Transformer to effectively process variable-length sparse windows and capture cross-window correlation, and employing a novel voxel diffusion technique to enhance the accuracy of 3D object detection with sparse features.

Apart from the two mainstreams mentioned above, there are also some methods that focus on how to learn effective spatial geometric representation directly from 3D point clouds. One of the representative works is PointPillars [21], which utilizes PointNet [38] to learn the representation of point clouds, organized in the form of vertical columns (i.e. pillars). The ability to operate at speeds higher than 60 Hz makes PointPillars one of the most widely used end-to-end object detection methods in mobile robotics, particularly in autonomous driving. As a summary of this section, as Zhao et al. [72] astutely observed, it is unnecessary to be limited to end-to-end or pipeline, combining the two may yield more competitive performance.

The remainder of this section details the segmentation-classification pipeline. First, a method named "adaptive clustering" developed by the author is presented. Subsequently, an open-source suite for point cloud segmentation evaluation is introduced to facilitate performance comparisons between different methods. Next, several hand-crafted features for object classification are illustrated, followed by a description of how these features are used to train two distinct classifiers. Finally, a multi-target tracker for point cloud data is introduced.

3.3.1 Point Cloud Segmentation

Point cloud segmentation methods can be broadly classified according to the AI taxonomy into rule-based (symbolism), traditional machine learning-based (statistics), and deep learning-based (connectionism) approaches. Rule-based ones typically segment point clouds based on their geometric features, intensity, surface normals and other information. These methods offer advantages such as high computational efficiency, robustness, and good interpretability. However, they struggle with complex scenes and are susceptible to occlusion and noise. Traditional machine learning-based methods extract shallow point cloud features and segment them using data-driven models. These methods share similar advantages and limitations with rule-based approaches. Deep learning-based methods, such as the previously mentioned PointNet [38], learn deep, abstract feature representations from point cloud data through deep neural networks. While these methods overcome some limitations of the former two categories, they require substantial training data, are prone to overfitting, and currently lack interpretability, which are challenges that require further research. Consequently, combining rule-based and deep learning-based methods remains a promising direction for addressing practical problems. The clustering methods detailed in the remainder of this chapter are rule-based.

3.3.1.1 Adaptive Clustering

The input and output point cloud data representation of the adaptive clustering method follows Eq. 3.3. The first step of the method is to remove the points representing the ground from the point cloud, since they are not the points of interest and are usually

3.3 Object Detection in Point Clouds

connected to various objects (since most of the objects are grounded), causing a huge hassle in segmenting them. To do this, a thresholding method is used:

$$P^* = \{p_i \in P \mid z_i < threshold\} \quad (3.4)$$

which means that all points below a preset distance threshold in the vertical direction (here represented by the z-axis) are eliminated. It is important to note that Eq. 3.4 is defined within the sensor coordinate system, where the negative z-axis points downwards. As is easy to imagine, this one-size-fits-all approach has both benefits and limitations: the former includes simple implementation and high computational efficiency, while the latter includes two assumptions to ensure the performance of the method: a flat ground and the z-axis of the sensor being perpendicular to the ground. These assumptions can be relaxed by employing the local convexity criterion [33].

The second step of the adaptive clustering method is to segment the remaining point cloud into non-overlapping clusters:

$$C_i \cap C_j = \emptyset, \text{ for } i \neq j \implies \min \|p_i - p_j\|_2 \geq d^* \quad (3.5)$$

where $C_i, C_j \subset P^*$ represents two such clusters, and d^* denotes a maximum imposed distance threshold. Equation 3.5 stipulates that if the minimum Euclidean distance between a set of points $p_i \in P^*$ and another set $p_j \in P^*$ is greater than the given threshold, then the points in p_i are set to belong to the cluster C_i, while the points in p_j are set to belong to the cluster C_j. An efficient implementation of this idea is introduced in [42]. Equation 3.5 works well for dense or structured point clouds but suffers from sparse or unstructured ones. The specific manifestation is that if d^* is too small, a single object may be divided into multiple clusters, and if it is too large, multiple objects may be merged into a single cluster (see Fig. 3.7). In addition, in terms of implementation, the computational cost of this method increases with the distance between points.

Therefore, further processing is required, and one approach to overcoming this performance limitation is to employ an adaptive d^*. This necessitates understanding the characteristics of point cloud data generated by typical mechanical 3D lidars. Due to their physical structure (see Sect. 3.2.2), these lidars produce point clouds with high horizontal resolution and comparatively low vertical resolution, with point density decreasing as distance increases. A representative example is shown in Fig. 3.8, which illustrates the point cloud generated by a 16-layer lidar scanning a human at varying distances. This lidar has a horizontal resolution of 0.1° and a vertical resolution of 2°. As the distance increases, the vertical spacing between points becomes significantly more pronounced compared to the horizontal spacing.

Therefore, it is a straightforward idea to adapt the threshold d^* linearly with respect to the scan distance:

$$d^* = 2 \cdot r \cdot \tan \frac{\Theta}{2} \quad (3.6)$$

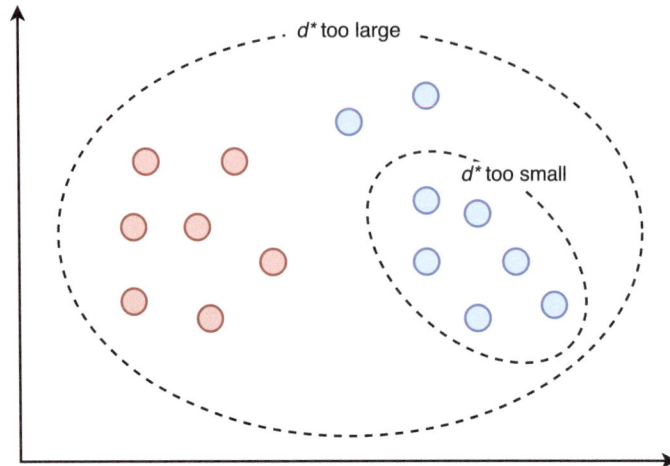

Fig. 3.7 Point cloud of two distinct objects (red and blue). Different values of d^* lead to different clustering results

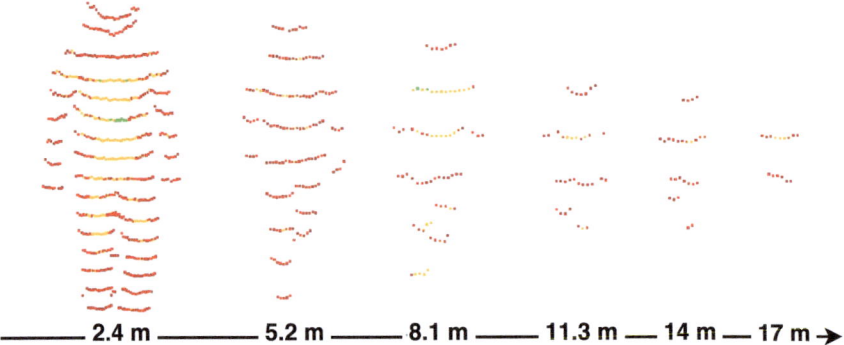

Fig. 3.8 Example of 3D lidar human point clouds at different distances. The further the person is from the sensor, the sparser is the corresponding point cloud

where r represents the scan distance and Θ is the lower resolution, e.g., for the aforementioned 16-layer lidar, $\Theta = 2°$. In the implementation of Eq. 3.6, an issue that needs to be considered is which points in P^* should be clustered using the same d^* value. By observing the morphology of ground data generated by 3D lidar and inspired by water ripples, a sensor-centered nested ring point cloud data segmentation method is proposed (see Fig. 3.9). The formal description is as follows. Consider a set of values d_i^* at fixed intervals Δd, where $d_{i+1}^* = d_i^* + \Delta d$. For each of them, the maximum cluster detection range r_i is calculated using the inverse of Eq. 3.6 and the corresponding radius $R_i = \lfloor r_i \rfloor$ is determined, where R_0 is the center of the sensor. The width of a region with constant threshold d_i^* is $l_i = R_i - R_{i-1}$. Therefore, the points in each ring are clustered using the same d_i^*. Moreover, the scale of d^* needs

3.3 Object Detection in Point Clouds

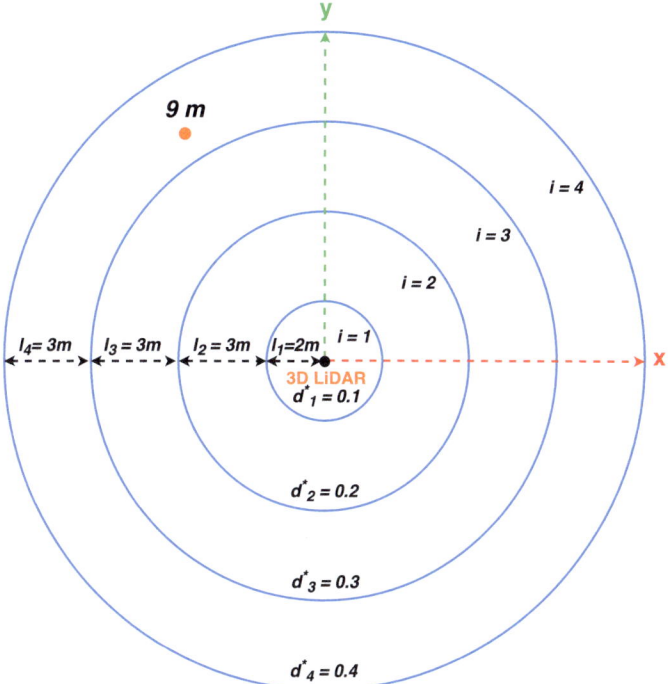

Fig. 3.9 Different nested regions are clustered using different d^* values. In this example, the cluster (orange point) at a distance of 9 m from the sensor is located on the 4th circular region, and its clustering threshold is $d_4^* = 0.4$ m

to be considered during actual operation. For example, a good practice for clustering points representing humans generated by the 16-layer lidar uses $\Delta d = 0.1$ m to obtain circular regions with width 2 to 3 m.

It is worth mentioning that filters can be used to further optimize the performance of the algorithm for specific tasks. For instance, to make human detection and tracking more effective, a volume-based filter can be used to filter out clusters that are too large or too small [56, 57]:

$$\overline{C} = \{C_i \mid 0.2 \leq w_i \leq 1,\ 0.2 \leq d_i \leq 1,\ 0.2 \leq h_i \leq 2\} \quad (3.7)$$

where w_i, d_i, and h_i are the width, depth, and height (in meters) of the bounding volume containing C_i, respectively. The adaptive clustering method is implemented based on a k-d tree and has a time complexity of $\mathcal{O}(\log n)$.

3.3.1.2 Benchmarking

In order to effectively evaluate various methods and gain insights into their strengths and weaknesses, the *LiDAR Point Cloud Clustering Benchmark Suite*[3] [68] can be used. It provides evaluations of five open-source methods on three reproduced open datasets as community baselines. For the former, in addition to the adaptive clustering method, it also includes:

- Run-based clustering [71], which consists of two steps. First, points representing the ground are extracted iteratively using deterministically assigned seed points. Then the remaining non-ground points are clustered using a two-run connected component labeling technique in binary images.
- Depth clustering [7], which is a fast and computationally inexpensive method that first converts 3D lidar scans into 2D range images and then segments the latter.
- Euclidean clustering [42], which clusters points by directly calculating the L_2 distance between any two points in 3D space.
- Autoware clustering [18], which is a modified version of Euclidean clustering. It first implicitly projects the points onto the 2D (x-y) plane and then segments them based on the L_2 distance.

Three open datasets were re-annotated and the five open source methods mentioned above were run on them. These datasets were collected outdoors using three different lidars, including the L-CAS dataset collected using a Velodyne VLP-16 [57], the EU long-term dataset collected using a Velodyne HDL-32E [61], and the KITTI dataset collected using a Velodyne HDL-64E [15]. The L-CAS dataset was collected in a parking lot with a stationary robot and contains two fully labeled pedestrians with no occlusions or truncations in the samples. The other two datasets were collected in urban road environments with a human-driven instrumented vehicle. The EU Long-term dataset provides labels for cars near a roundabout, while the KITTI dataset includes annotations for pedestrians, cyclists, and various vehicle types.

Accurate performance evaluation hinges on high-quality ground truth annotations. Therefore, the following steps were taken to ensure the quality of the annotations for each dataset:

- L-CAS dataset: The accuracy of the existing annotations was improved.
- EU Long-term dataset: 200 consecutive frames were extracted from the roundabout data collected on April 12, 2019. All vehicles within these frames were meticulously annotated.
- KITTI dataset: To address limitations in the original KITTI annotations, 200 frames were randomly selected from the 3D object detection data and re-annotated. The original bounding boxes, based on RGB image projection, estimate the full vehicle size and might not objectively reflect the clustering method's performance.

Additionally, for the EU Long-term and KITTI datasets, a ray ground filter [16] was employed to remove ground points instead of a simple z-axis threshold. This approach

---

[3] https://github.com/cavayangtao/lidar_clustering_bench.

3.3 Object Detection in Point Clouds

Table 3.2 Clustering accuracy of different methods on different datasets

Approach	Parameters				Precision (best in bold)		
	Ground removal	Min/Max points	Clustering θ		L-CAS (%)	EU long-term (%)	KITTI (%)
Run clustering [71]	$Params_{GPF}$	2/2.2 million	$Params_{SLR}$		37.63	**35.97**	29.25
Depth clustering [7]	7°	5/2.2 million	10°		14.61	28.72	**42.69**
Euclidean clustering [43]	− 0.8 m/− 1.25 m/− 1.5 m	5/2.2 million	0.75 m		39.26	14.78	30.63
Autoware clustering [18]	− 0.8 m/− 1.25 m/− 1.5 m	5/2.2 million	0.75 m		50.68	34.00	32.15
Adaptive clustering [57]	− 0.8 m/− 1.25 m/− 1.5 m	5/2.2 million	Adaptive		**62.38**	32.99	33.24
Euclidean clustering [43]	Ray ground filter	5/2.2 million	0.75 m		22.19	**71.12**	67.16
Autoware clustering [18]	Ray ground filter	5/2.2 million	0.75 m		27.20	62.13	**70.86**
Adaptive clustering [57]	Ray ground filter	5/2.2 million	Adaptive		31.81	37.13	20.28

$Params_{GPF} = \{N_{segs} = 3, N_{iter} = 3, N_{LPR} = 20, Th_{seeds} = 0.4\,\text{m}, Th_{dist} = 0.2\,\text{m}\}$
$Params_{SLR} = \{Th_{run} = 0.5\,\text{m}, Th_{merge} = 1\,\text{m}\}$

improves the completeness of the annotated bounding boxes. All annotations were consistently performed using the L-CAS 3D Point Cloud Annotation Tool[4] [57].

Furthermore, to ensure compatibility with run clustering [71] and depth clustering [7], which rely on ring information, the missing ring number (corresponding to the 3D lidar's scanning layer) in the KITTI point cloud data was estimated using the following equation:

$$ring = \left\lfloor n \times \frac{\arcsin(\frac{z}{\sqrt{x^2+y^2+z^2}}) + FoV_{down}}{FoV} \right\rfloor \quad (3.8)$$

where n is the number of lidar layers, (x, y, z) are the coordinates of a laser point, FoV is the lidar's vertical field of view, and FoV_{down} is the vertical field of view below 0°. The resulting ring number is rounded to the nearest integer.

The 3D Intersection over Union (3D IoU) between clustered and ground truth bounding boxes serves as the metric for evaluating the clustering accuracy of each method. The benchmarking results are presented in Table 3.2. All methods were executed using the parameters specified in their respective publications. When a parameter value was not explicitly defined, the optimal value determined through experimentation was used. As shown in the table, the adaptive clustering method

---

[4] https://github.com/yzrobot/cloud_annotation_tool.

achieves the highest performance on the L-CAS dataset, thanks to its direct computation of the Euclidean distance between different points in 3D space. Conversely, the depth clustering method's performance is negatively impacted by edge cases, particularly when objects are in close proximity and background objects are larger than foreground one. The run clustering and depth clustering methods demonstrate superior performance on the EU Long-term and KITTI datasets, respectively, due to their increased robustness to uneven and sloped road surfaces.

To further analyze the performance of the evaluated methods, additional experiments were conducted to investigate the impact of ground removal. As previously mentioned, while threshold-based ground removal is often employed to meet the real-time constraints of robotic systems, it relies on the assumption of a flat ground surface. Therefore, it is pertinent to investigate whether more sophisticated ground removal techniques can enhance the performance of related clustering methods. As shown in the final three rows of Table 3.2, applying the ray ground filter [16] generally improves the clustering performance of the original threshold-based methods on the EU Long-term and KITTI datasets, but leads to a decrease in performance on the L-CAS dataset. This decline is attributed to the filter inadvertently removing portions of pedestrian feet, resulting in clustered bounding boxes that are smaller than the ground truth in most cases.

On the other hand, the runtime of each clustering method was evaluated. The experiments were performed on Ubuntu 18.04 LTS (64-bit) and ROS Melodic, using an Intel i7-7700HQ processor (only one core is used), 16 GB of memory, and no GPU was used. The experimental results are shown in Fig. 3.10. It can be seen that the processing time of all methods is proportional to the number of points contained in the point cloud, with the 16-layer lidar data exhibiting the shortest processing time and the 64-layer data the longest. Depth clustering demonstrates a significant performance advantage due to its dimensionality reduction of the point clouds and its optimized implementation. Run clustering also exhibits competitive runtime, as it effectively leverages prior knowledge such as ring information, thus avoiding full point cloud traversal. The remaining three methods are implemented using the k-d tree provided by PCL, resulting in an average time complexity of $\mathcal{O}(kn \log n)$, which is comparatively time-consuming. Among these three, adaptive clustering is the fastest, owing to its ring-based point cloud partitioning, which reduces the k-d tree search space. Autoware clustering outperforms Euclidean clustering for larger point clouds, as it performs clustering in the 2D plane after projecting the 3D point cloud, whereas Euclidean clustering performs the tree search directly in 3D space.

3.3.2 Object Classification

Object classification in point clouds involves assigning semantic labels to individual points or groups of points. This section focuses on hand-crafted features and traditional machine learning models.

3.3 Object Detection in Point Clouds

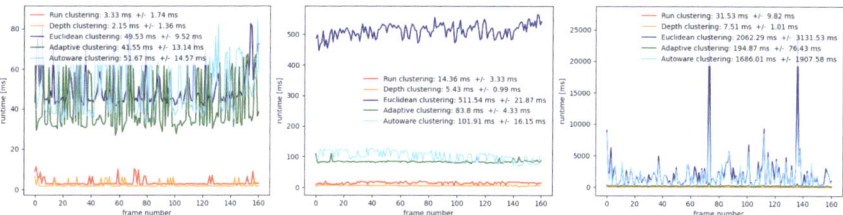

Fig. 3.10 Runtime performance of the evaluated clustering methods

Table 3.3 Handcrafted features used for object classification

Feature	Description	Dimension
f_1	Number of points included in the cluster	1
f_2	Minimum cluster distance from the sensor	1
f_3	3D covariance matrix of the cluster	6
f_4	Normalized moment of inertia tensor	6
f_5	2D covariance matrix in 3 zones including the upper half, the left and right lower halves	9
f_6	The normalized 2D histogram for the main PCA plane	98
f_7	The normalized 2D histogram for the secondary PCA plane	45
f_8	Slice feature for the cluster	20
f_9	Reflection intensity's distribution (mean, standard deviation and normalized 1D histogram)	27
f_{10}	Distance from the centroid of each slice to the sensor	10

3.3.2.1 Hand-Crafted Features

Table 3.3 summarizes commonly used hand-crafted features for point clouds. Among them, $\{f_1, \ldots, f_7\}$ were introduced by Navarro-Serment et al. [35], while $\{f_8, f_9\}$ were proposed by Kidono et al. [20]. Due to the "learn quickly and deploy immediately" requirement of Robot Online Learning (ROL) [60], a balance between feature completeness and computational efficiency is necessary. Specifically, based on analyses in [20] and [57], the feature set $\{f_1, \ldots, f_4, f_8, f_9\}$ has been used for pedestrian, cyclist, and car detection in mobile robotics [56, 59, 63–65]. The feature f_{10}, termed "slice distance", was proposed by Yan et al. [57] and has been shown to improve pedestrian detection accuracy. To understand this feature, it is necessary to first explain the "slice feature" f_8. This feature divides the 3D points within a cluster into 10 equal-height slices and calculates the length and width of each slice (see Fig. 3.11):

$$f_8 = \{L_j, W_j \mid j = 1, \ldots, 10\} \tag{3.9}$$

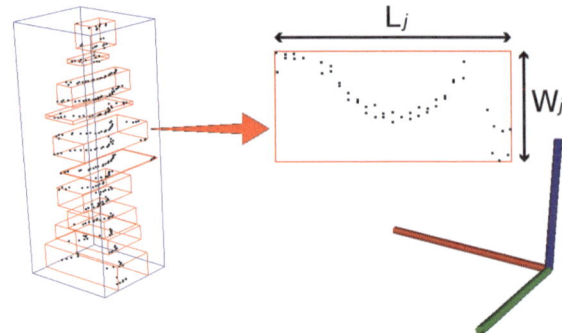

Fig. 3.11 The slice feature proposed by Kidono et al. [20]

Therefore, the slice distance aims to relate the distance measurement of the sensor to the 3D shape of the human body, which is particularly useful for classifying sparse point clouds at long ranges. It is calculated as the Euclidean distance of each slice centroid to the origin:

$$f_{10} = \{\|c_i\|_2 \mid c_i = (x_i, y_i, z_i) \in \mathbb{R}^3, i = 1, \ldots, 10\} \quad (3.10)$$

where c_i is the centroid of the i-th slice:

$$c_i = \frac{1}{|C_i|} \sum_{p_i \in C_i} p_i \quad (3.11)$$

where C_i is the set of points in the i-th slice.

3.3.2.2 Learning Models

Pedestrian classification in 3D point clouds is typically a nonlinear classification problem. This is because 3D point cloud data is inherently nonlinear, and the decision boundary between pedestrian and non-pedestrian points is unlikely to be a simple hyperplane. As illustrated in Fig. 3.12, the shapes of pedestrians can vary significantly due to differences in pose, size, and orientation, as well as point cloud sparsity, occlusion, and background complexity. Commonly used nonlinear classifiers include Support Vector Machine (SVM), Random Forest (RF) and DNN. SVM has a strong mathematical foundation and perform well with limited data, making it suitable for ROL [56, 56, 57], as discussed further in Chap. 4. RF can satisfy the ROL requirements for faster model training and online adaptability in long-term, cross-environment deployments [63–65]. While DNN is a promising model, its current computational demands and challenges in real-time model updating place it outside the scope of this section.

3.3 Object Detection in Point Clouds

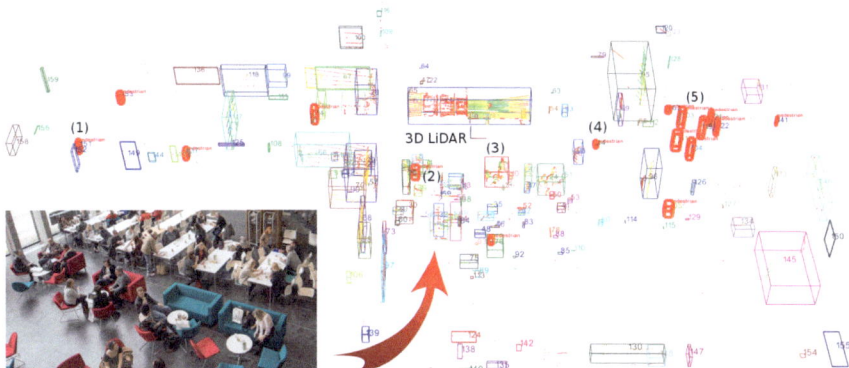

Fig. 3.12 Example of annotated point clouds generated in an academic building [57]. Ground and ceiling points have been removed. Each bounding box represents a cluster, and red boxes are humans. The image contains some typical annotation challenges: (1) A human sample at 25 m from the sensor consisting of only five points; (2) A cluster of a seated person, merged with a chair; (3) A cluster of two people sitting at a round table, which clearly could not be labeled as human; (4) A human head only and the rest of the body is occluded; (5) A set of clusters of people with many occlusions

Next, we detail the implementation of SVM-based [56, 56, 57] and RF-based [63–65] classification. For the SVM approach, in each ROL iteration, a binary SVM classifier is trained using the features described earlier to distinguish between human and non-human clusters. LIBSVM [10] is used for training. The ratio of positive to negative training samples is 1 : 1, and all data values are scaled to $[-1, 1]$. A Gaussian Radial Basis Function (RBF) kernel [19] is employed, and the SVM outputs class probabilities. The original implementation retrains the classifier from scratch in each iteration, using all accumulated training examples. Consequently, the training time is proportional to the number of samples, ranging from sub-millisecond to several minutes. However, the training process can be decoupled from ROL (e.g., using independent threads [29, 70]) or accelerated by optimizing the k-fold cross-validation used for hyperparameter tuning.

For RF-based classification, an online variant, named Online Random Forest (ORF) [44], is employed. The latter combines the ideas of online bagging and extreme RF, and proposes an online decision tree growth strategy that allows its performance with streaming data to converge to that of offline RF with batch data. Specifically, a tree node is split based on two criteria: 1) whether the node contains enough samples for robust statistics, and 2) whether the split provides sufficient classification gain. These two conditions are formalized as:

$$|R_j| > \alpha \quad \wedge \quad \exists s \in S : \Delta L(R_j, s) > \beta \tag{3.12}$$

where α is the minimum number of samples required in a node before splitting, β is the minimum gain required for a split, R_j represents a decision node, and $\Delta L(R_j, s)$ is the gain of node j with respect to test s, calculated as:

$$\Delta L(R_j, s) = L(R_j) - \frac{|R_{jls}|}{|R_j|} L(R_{jls}) - \frac{|R_{jrs}|}{|R_j|} L(R_{jrs}) \tag{3.13}$$

where R_{jls} and R_{jrs} are the left and right child nodes resulting from split s. The split with the highest gain is chosen for the node:

$$s_j = \underset{s \in S}{\operatorname{argmax}} \Delta L(R_j, s) \tag{3.14}$$

Building upon the original ORF, we have added support for streaming data, mini-batch learning, and real-time model storage. Further details are available in the released code.[5]

3.4 Multi-target Tracking in Point Clouds

In mobile robotics, detecting and tracking moving objects is key to achieving useful and safe robot behaviors. Similar to object detection (cf. Sect. 3.3), multi-target tracking also has both end-to-end and pipeline approaches. The former directly processes raw sensor data and outputs the motion trajectory of each target. It combines two tasks, object detection and object tracking, into a single model, typically based on DNN [11]. The latter is usually referred to as the tracking-by-detection paradigm, as illustrated in Fig. 3.13. Specifically, objects are perceived by one or more sensors, whose respective data (e.g., images or distance measurements) are processed by software algorithms and produce position estimates of the objects relative to the robot, such as orientation and/or distance, termed "observations". Then in the tracking phase, the observations are associated with previous or new object motion estimates via some data association algorithms and are ultimately used by one or more estimators (e.g., Kalman filter, particle filter, etc.) to update these estimates.

Methodologically, the multi-target tracking process consists of two stages (as shown in Fig. 3.14): data association and state estimation. In the first stage, new observations from object detectors are compared with previous predictions from the tracker and either associated with the latter or discarded. In the second stage, the state estimates of the objects are updated based on their associated observations.

---

[5] https://github.com/RuiYang-1010/efficient_online_learning.

3.4 Multi-target Tracking in Point Clouds

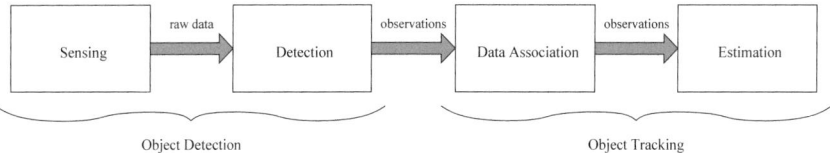

Fig. 3.13 General architecture of a tracking-by-detection system [3]

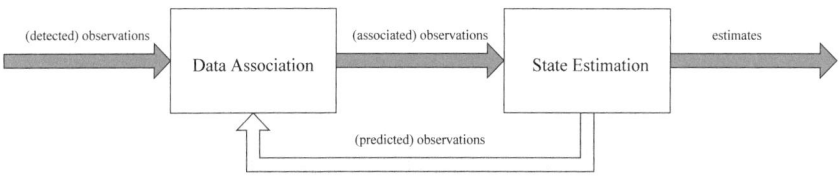

Fig. 3.14 An overview of the tracking process, including data association and state estimation [3]

3.4.1 Data Association

Two common data association algorithms are Global Nearest Neighbor (GNN) [4, 27, 34, 36] and Joint Probabilistic Data Association (JPDA) [2, 6, 46]. The former associates new observations by matching them with their closest predicted ones. The distance is usually a statistical measure that takes into account the uncertainty of the track estimates and sensor readings (e.g., Mahalanobis distance). GNN is a "one-to-one" association algorithm, where each target can be associated with only one observation, and each observation can be generated by only one target. Observations that have not been associated with any target at the end of the procedure are usually used for the creation of new tracks. The advantage of this method is that it is fast and can therefore handle a large number of targets and observations. The limitation is that it struggles to handle occlusions.

For each observation and each tracked target, JPDA calculates their association probability based on the distance between the observation and the target's predicted state and the target's survival probability. Different from data association of GNN, JPDA employs a many-to-one/one-to-many approach, where several observations can be generated by the same target, and several targets can give origin to the same observation. This increased flexibility makes JPDA more robust to certain scenarios, such as short occlusions, but also computationally more expensive than GNN. Therefore, we suggest using the JPDA method for small-scale multi-target tracking and the GNN method for large-scale multi-target tracking.

3.4.2 State Estimation

The Unscented Kalman Filter (UKF) is an effective method for target state estimation. For human tracking in point clouds, the prediction step can be based on the following constant velocity model [23]:

$$\begin{cases} x_k = x_{k-1} + \Delta t \, \dot{x}_{k-1} \\ \dot{x}_k = \dot{x}_{k-1} \\ y_k = y_{k-1} + \Delta t \, \dot{y}_{k-1} \\ \dot{y}_k = \dot{y}_{k-1} \end{cases} \quad (3.15)$$

where x_k and y_k are the Cartesian coordinates of the target at time t_k, $\dot{x}_k$ and $\dot{y}_k$ are their respective velocities, and $\Delta t = t_k - t_{k-1}$. The position of the cluster representing the human is calculated by projecting its centroid c_i (see Eq. 3.11) onto the xy-plane:

$$c'_i = \lambda \cdot c_i, \quad \lambda = (1, 1, 0) \quad (3.16)$$

The update step of the estimation then uses a 2D polar observation model to represent the position of the cluster:

$$\begin{cases} \theta_k = tan^{-1}(y_k/x_k) \\ \gamma_k = \sqrt{x_k^2 + y_k^2} \end{cases} \quad (3.17)$$

where θ_k and γ_k represent the cluster's azimuth and distance with respect to the sensor, respectively. Note that for simplicity, noise and coordinate transformations, including those associated with the robot motion, are omitted in the above equation. The choice of the polar observation model over a Cartesian observation model for 3D lidar is driven by the data and noise characteristics of this sensor. Its measurements are range values at regular angular intervals, and the noise is directional. The nonlinearity of the model also leads to the adoption of UKF, which outperforms the standard Extended Kalman Filter (EKF). The estimation framework is shown to be an effective solution for human tracking with mobile robots [5, 27]. For further details on track management (including initialization, maintenance, and deletion) and possible applications, see [4, 5, 13].

Finally, the process noise covariance matrix Q for the prediction model and the measurement noise covariance matrix R for the observation model are defined as:

$$Q = \begin{bmatrix} \frac{\Delta t^4}{4}\sigma_x^2 & \frac{\Delta t^3}{2}\sigma_x^2 & 0 & 0 \\ \frac{\Delta t^3}{2}\sigma_x^2 & \Delta t^2 \sigma_x^2 & 0 & 0 \\ 0 & 0 & \frac{\Delta t^4}{4}\sigma_y^2 & \frac{\Delta t^3}{2}\sigma_y^2 \\ 0 & 0 & \frac{\Delta t^3}{2}\sigma_y^2 & \Delta t^2 \sigma_y^2 \end{bmatrix} \quad R = \begin{bmatrix} \sigma_\theta^2 & 0 \\ 0 & \sigma_\gamma^2 \end{bmatrix} \quad (3.18)$$

where σ_x, σ_y, σ_θ, and σ_γ represent the standard deviations of the respective noise components and are empirically determined to optimize human tracking performance for the given robot platform.

3.5 Conclusion

This chapter explored research in robot perception, specifically focusing on 3D lidar-based human detection and tracking. We began by summarizing fundamental aspects of 3D lidar, covering its ranging principles, scanning architectures, physical characteristics, data representation, and practical applications within industry. Next, a pipeline-based object detection approach was presented, encompassing point cloud segmentation and object classification. For the former, the "adaptive clustering" method was detailed, alongside the "LiDAR Point Cloud Clustering Benchmark Suite", designed to enable fair performance comparisons across various techniques. For the latter, we initially examined hand-crafted features for human classification, then extended this to include the differentiation of cars, cyclists, and pedestrians, employing SVM and RF as learning models. Finally, the multi-target tracking system was described, with particular emphasis on the state estimation step, specifically adapted for tracking individuals within point cloud data.

Generally speaking, human detection based on 3D lidar is one of the key technologies in the fields of mobile robotics, autonomous driving, smart security and more. In recent years, with the development of hardware and AI technology, research on this topic has made rapid progress. However, several challenges still exist. First, the prices of the high-performance 3D lidars currently in mass production are not yet generally affordable. This reflects, from one aspect, the importance of datasets in promoting the development of related technologies. Fortunately, there are already numerous datasets available in the community (see Sect. 2.4). In the future, we hope that more high-quality datasets for different scenes and scenarios will appear, particularly those based on multi-sensor perception systems. Second, as mentioned earlier, the point clouds generated by most current 3D lidars are sparse, making it difficult to extract effective features for small and distant objects. This not only puts forward the demand for the continued development of hardware, but also creates space for the joint use of 3D lidar with other modal sensors such as cameras. Finally, due to its physical characteristics, the performance of lidar is sensitive to adverse weather conditions such as rain, fog, and snow. How to model and reduce the noise caused by water droplets in the air is a research direction worthy of in-depth study [66, 67, 69].

References

1. Aptiv, A.U.D.I., Baidu, B.M.W.: Continental, Daimler, FCA, HERE, Infineon, Intel, and Volkswagen: safety first for automated driving. Tech. rep, Withe Paper (2019)
2. Bar-Shalom, Y., Li, X.: Multitarget-Multisensor Tracking: Principles and Techniques. Y. Bar-Shalom (1995)
3. Bellotto, N., Cosar, S., Yan, Z.: Human detection and tracking. In: M.H. Ang, O. Khatib, B. Siciliano (eds.) Encyclopedia of Robotics, pp. 1–10. Springer (2018)
4. Bellotto, N., Hu, H.: Multisensor-based human detection and tracking for mobile service robots. IEEE Trans. Syst., Man, Cybern.—Part B **39**(1), 167–181 (2009). https://doi.org/10.1109/TSMCB.2008.2004050
5. Bellotto, N., Hu, H.: Computationally efficient solutions for tracking people with a mobile robot: an experimental evaluation of Bayesian filters. Autonom. Robots **28**(4), 425–438 (2010). https://doi.org/10.1007/S10514-009-9167-2
6. Bennewitz, M., Burgard, W., Cielniak, G., Thrun, S.: Learning motion patterns of people for compliant robot motion. Int. J. Robot. Res. **24**(1), 31–48 (2005). https://doi.org/10.1177/0278364904048962
7. Bogoslavskyi, I., Stachniss, C.: Fast range image-based segmentation of sparse 3d laser scans for online operation. In: IEEE/RSJ International Conference on Intelligent Robots and Systems (IROS), pp. 163–169 (2016). https://doi.org/10.1109/IROS.2016.7759050
8. Broughton, G., Janota, J., Blaha, J., Yan, Z., Krajnik, T.: Bootstrapped learning for car detection in planar lidars. In: Proceedings of the 2022 ACM/SIGAPP Symposium on Applied Computing (SAC), pp. 758–765. Virtual Conference (2022)
9. Caesar, H., Bankiti, V., Lang, A.H., Vora, S., Liong, V.E., Xu, Q., Krishnan, A., Pan, Y., Baldan, G., Beijbom, O.: nuscenes: a multimodal dataset for autonomous driving. In: IEEE/CVF Conference on Computer Vision and Pattern Recognition (CVPR), pp. 11618–11628 (2020). https://doi.org/10.1109/CVPR42600.2020.01164
10. Chang, C., Lin, C.: LIBSVM: a library for support vector machines. ACM Trans. Intell. Syst. Technol. **2**(3), 1–27 (2011). https://doi.org/10.1145/1961189.1961199
11. Dequaire, J., Ondruska, P., Rao, D., Wang, D.Z., Posner, I.: Deep tracking in the wild: end-to-end tracking using recurrent neural networks. Int. J. Robot. Res. **37**(4–5), 492–512 (2018). https://doi.org/10.1177/0278364917710543
12. Dewan, A., Caselitz, T., Tipaldi, G.D., Burgard, W.: Motion-based detection and tracking in 3d lidar scans. In: D. Kragic, A. Bicchi, A.D. Luca (eds.) IEEE International Conference on Robotics and Automation (ICRA), pp. 4508–4513 (2016). https://doi.org/10.1109/ICRA.2016.7487649
13. Dondrup, C., Bellotto, N., Jovan, F., Hanheide, M.: Real-time multisensor people tracking for human-robot spatial interaction. In: ICRA Workshop on Machine Learning for Social Robotics, pp. 1–6 (2015)
14. Duckett, T., Pearson, S., Blackmore, S., Grieve, B.: Agricultural robotics: the future of robotic agriculture. CoRR https://arxiv.org/abs/1806.06762 (2018)
15. Geiger, A., Lenz, P., Urtasun, R.: Are we ready for autonomous driving? The KITTI vision benchmark suite. In: IEEE/CVF Conference on Computer Vision and Pattern Recognition (CVPR), pp. 3354–3361 (2012). https://doi.org/10.1109/CVPR.2012.6248074
16. Himmelsbach, M., von Hundelshausen, F., Wünsche, H.: Fast segmentation of 3d point clouds for ground vehicles. In: IEEE Intelligent Vehicles Symposium (IV), pp. 560–565 (2010). https://doi.org/10.1109/IVS.2010.5548059
17. Jeong, J., Cho, Y., Shin, Y., Roh, H.C., Kim, A.: Complex urban dataset with multi-level sensors from highly diverse urban environments. Int. J. Robot. Res. **38**(6) (2019). https://doi.org/10.1177/0278364919843996
18. Kato, S., Tokunaga, S., Maruyama, Y., Maeda, S., Hirabayashi, M., Kitsukawa, Y., Monrroy, A., Ando, T., Fujii, Y., Azumi, T.: Autoware on board: enabling autonomous vehicles with embedded systems. In: 9th ACM/IEEE International Conference on Cyber-Physical Systems (ICCPS), pp. 287–296 (2018). https://doi.org/10.1109/ICCPS.2018.00035

References

19. Keerthi, S.S., Lin, C.: Asymptotic behaviors of support vector machines with Gaussian kernel. Neural Comput. **15**(7), 1667–1689 (2003). https://doi.org/10.1162/089976603321891855
20. Kidono, K., Miyasaka, T., Watanabe, A., Naito, T., Miura, J.: Pedestrian recognition using high-definition LIDAR. In: IEEE Intelligent Vehicles Symposium (IV), pp. 405–410 (2011). https://doi.org/10.1109/IVS.2011.5940433
21. Lang, A.H., Vora, S., Caesar, H., Zhou, L., Yang, J., Beijbom, O.: Pointpillars: fast encoders for object detection from point clouds. In: IEEE Conference on Computer Vision and Pattern Recognition (CVPR), pp. 12697–12705 (2019). https://doi.org/10.1109/CVPR.2019.01298
22. Lekner, J., Dorf, M.C.: Why some things are darker when wet. Appl. Opt. **27**(7), 1278–1280 (1988)
23. Li, X.R., Jilkov, V.P.: Survey of maneuvering target tracking. part I. dynamic models. IEEE Trans. Aerosp. Electron. Syst. **39**(4), 1333–1364 (2003)
24. Li, Y., Duthon, P., Colomb, M., Ibañez-Guzmán, J.: What happens for a ToF LiDAR in fog? IEEE Trans. Intell. Transp. Syst. **22**(11), 6670–6681 (2021). https://doi.org/10.1109/TITS.2020.2998077
25. Li, Y., Ibañez-Guzmán, J.: Lidar for autonomous driving: the principles, challenges, and trends for automotive lidar and perception systems. IEEE Signal Process. Mag. **37**(4), 50–61 (2020). https://doi.org/10.1109/MSP.2020.2973615
26. Lin, T., Goyal, P., Girshick, R.B., He, K., Dollár, P.: Focal loss for dense object detection. In: IEEE International Conference on Computer Vision (ICCV), pp. 2999–3007 (2017). https://doi.org/10.1109/ICCV.2017.324
27. Linder, T., Breuers, S., Leibe, B., Arras, K.O.: On multi-modal people tracking from mobile platforms in very crowded and dynamic environments. In: IEEE International Conference on Robotics and Automation (ICRA), pp. 5512–5519 (2016). https://doi.org/10.1109/ICRA.2016.7487766
28. Linder, T., Vaskevicius, N., Schirmer, R., Arras, K.O.: Cross-modal analysis of human detection for robotics: an industrial case study. In: IEEE/RSJ International Conference on Intelligent Robots and Systems (IROS), pp. 971–978 (2021). https://doi.org/10.1109/IROS51168.2021.9636158
29. Liu, B., Yao, D., Yang, R., Yan, Z., Yang, T.: Semi-supervised online continual learning for 3d object detection in mobile robotics. J. Intell. Robotic Syst. **110**(4), 1–16 (2024)
30. Loey, M., Manogaran, G., Taha, M.H.N., Khalifa, N.E.M.: A hybrid deep transfer learning model with machine learning methods for face mask detection in the era of the covid-19 pandemic. Measurement **167**, 108288 (2021)
31. Maddern, W., Pascoe, G., Linegar, C., Newman, P.: 1 year, 1000 km: the oxford RobotCar dataset. Int. J. Robot. Res. **36**(1), 3–15 (2017). https://doi.org/10.1177/0278364916679498
32. Montemerlo, M., Becker, J., Bhat, S., Dahlkamp, H., Dolgov, D., Ettinger, S., Hähnel, D., Hilden, T., Hoffmann, G., Huhnke, B., Johnston, D., Klumpp, S., Langer, D., Levandowski, A., Levinson, J., Marcil, J., Orenstein, D., Paefgen, J., Penny, I., Petrovskaya, A., Pflueger, M., Stanek, G., Stavens, D., Vogt, A., Thrun, S.: Junior: the Stanford entry in the urban challenge. J. Field Robot. **25**(9), 569–597 (2008). https://doi.org/10.1002/ROB.20258
33. Moosmann, F., Pink, O., Stiller, C.: Segmentation of 3d lidar data in non-flat urban environments using a local convexity criterion. In: 2009 IEEE Intelligent Vehicles Symposium (IV), pp. 1931–1587 (2009)
34. Munaro, M., Menegatti, E.: Fast RGB-D people tracking for service robots. Auton. Robots **37**(3), 227–242 (2014). https://doi.org/10.1007/S10514-014-9385-0
35. Navarro-Serment, L.E., Mertz, C., Hebert, M.: Pedestrian detection and tracking using three-dimensional LADAR data. Int. J. Robot. Res. **29**(12), 1516–1528 (2010). https://doi.org/10.1177/0278364910370216
36. Pesenti Gritti, A., Tarabini, O., Guzzi, J., Caro, G.A.D., Caglioti, V., Gambardella, L.M., Giusti, A.: Kinect-based people detection and tracking from small-footprint ground robots. In: IEEE/RSJ International Conference on Intelligent Robots and Systems (IROS), pp. 4096–4103 (2014). https://doi.org/10.1109/IROS.2014.6943139

37. Pitropov, M., Garcia, D.E., Rebello, J., Smart, M., Wang, C., Czarnecki, K., Waslander, S.L.: Canadian adverse driving conditions dataset. Int. J. Robot. Res. **40**(4–5) (2021). https://doi.org/10.1177/0278364920979368
38. Qi, C.R., Su, H., Mo, K., Guibas, L.J.: Pointnet: deep learning on point sets for 3d classification and segmentation. In: IEEE Conference on Computer Vision and Pattern Recognition (CVPR), pp. 77–85 (2017). https://doi.org/10.1109/CVPR.2017.16
39. Rasshofer, R.H., Spies, M., Spies, H.: Influences of weather phenomena on automotive laser radar systems. Adv. Radio Sci. **9**, 49–60 (2011)
40. Redmon, J., Divvala, S.K., Girshick, R.B., Farhadi, A.: You only look once: unified, real-time object detection. In: IEEE Conference on Computer Vision and Pattern Recognition (CVPR), pp. 779–788 (2016). https://doi.org/10.1109/CVPR.2016.91
41. Roucek, T., Pecka, M., Cížek, P., Petrícek, T., Bayer, J., Salanský, V., Azayev, T., Hert, D., Petrlík, M., Báca, T., Spurný, V., Krátký, V., Petrácek, P., Baril, D., Vaidis, M., Kubelka, V., Pomerleau, F., Faigl, J., Zimmermann, K., Saska, M., Svoboda, T., Krajník, T.: System for multi-robotic exploration of underground environments CTU-CRAS-NORLAB in the DARPA subterranean challenge. Field Robot. **2**(1), 1779–1818 (2022). https://doi.org/10.55417/FR.2022055
42. Rusu, R.B.: Semantic 3D object maps for everyday manipulation in human living environments. Ph.D. thesis, Computer Science Department, Technische Universitaet Muenchen, Germany (2009)
43. Rusu, R.B., Cousins, S.: 3d is here: point cloud library (PCL). In: IEEE International Conference on Robotics and Automation (ICRA) (2011). https://doi.org/10.1109/ICRA.2011.5980567
44. Saffari, A., Leistner, C., Santner, J., Godec, M., Bischof, H.: On-line random forests. In: 12th IEEE International Conference on Computer Vision Workshops (ICCV Workshops), pp. 1393–1400 (2009). https://doi.org/10.1109/ICCVW.2009.5457447
45. Sathyamoorthy, A.J., Patel, U., Paul, M., Savle, Y., Manocha, D.: COVID surveillance robot: monitoring social distancing constraints in indoor scenarios. PLoS One **16**(12), e0259713 (2021)
46. Schulz, D., Burgard, W., Fox, D., Cremers, A.B.: People tracking with mobile robots using sample-based joint probabilistic data association filters. Int. J. Robot. Res. **22**(2), 99–116 (2003). https://doi.org/10.1177/0278364903022002002
47. Shackleton, J., Voorst, B.V., Hesch, J.A.: Tracking people with a 360-degree lidar. In: Seventh IEEE International Conference on Advanced Video and Signal Based Surveillance (AVSS), pp. 420–426 (2010). https://doi.org/10.1109/AVSS.2010.52
48. Simon, M., Milz, S., Amende, K., Gross, H.: Complex-YOLO: an Euler-region-proposal for real-time 3d object detection on point clouds. In: European Conference on Computer Vision Workshops (ECCV Workshops), vol. 11129, pp. 197–209 (2018). https://doi.org/10.1007/978-3-030-11009-3_11
49. Sualeh, M., Kim, G.: Dynamic multi-lidar based multiple object detection and tracking. Sensors **19**(6), 1474 (2019). https://doi.org/10.3390/S19061474
50. Sun, L., Taher, M., Wild, C., Zhao, C., Majer, F., Yan, Z., Krajnik, T., Prescott, T.J., Duckett, T.: Robust and long-term monocular teach-and-repeat navigation using a single-experience map. In: Proceedings of the 2021 IEEE/RSJ International Conference on Intelligent Robots and Systems (IROS), pp. 2635–2642. Prague, Czech Republic (2021)
51. Sun, P., Kretzschmar, H., Dotiwalla, X., Chouard, A., Patnaik, V., Tsui, P., Guo, J., Zhou, Y., Chai, Y., Caine, B., Vasudevan, V., Han, W., Ngiam, J., Zhao, H., Timofeev, A., Ettinger, S., Krivokon, M., Gao, A., Joshi, A., Zhang, Y., Shlens, J., Chen, Z., Anguelov, D.: Scalability in perception for autonomous driving: Waymo open dataset. In: IEEE/CVF Conference on Computer Vision and Pattern Recognition (CVPR), pp. 2443–2451 (2020). https://doi.org/10.1109/CVPR42600.2020.00252
52. Sun, P., Tan, M., Wang, W., Liu, C., Xia, F., Leng, Z., Anguelov, D.: Swformer: sparse window transformer for 3d object detection in point clouds. In: S. Avidan, G.J. Brostow, M. Cissé, G.M. Farinella, T. Hassner (eds.) 17th European Conference on Computer Vision (ECCV), vol. 13670, pp. 426–442 (2022). https://doi.org/10.1007/978-3-031-20080-9_25

53. Thrun, S., Burgard, W., Fox, D.: Probabilistic Robotics. MIT Press, Intelligent Robotics and Autonomous Agents (2005)
54. Vintr, T., Yan, Z., Duckett, T., Krajnik, T.: Spatio-temporal representation for long-term anticipation of human presence in service robotics. In: Proceedings of the 2019 IEEE International Conference on Robotics and Automation (ICRA), pp. 2620–2626. Montreal, Canada (2019)
55. Yan, Y., Mao, Y., Li, B.: SECOND: sparsely embedded convolutional detection. Sensors **18**(10), 3337 (2018). https://doi.org/10.3390/S18103337
56. Yan, Z., Duckett, T., Bellotto, N.: Online learning for human classification in 3D LiDAR-based tracking. In: Proceedings of the 2017 IEEE/RSJ International Conference on Intelligent Robots and Systems (IROS), pp. 864–871. Vancouver, Canada (2017)
57. Yan, Z., Duckett, T., Bellotto, N.: Online learning for 3d lidar-based human detection: experimental analysis of point cloud clustering and classification methods. Auton. Robots **44**(2), 147–164 (2020)
58. Yan, Z., Schreiberhuber, S., Halmetschlager, G., Duckett, T., Vincze, M., Bellotto, N.: Robot perception of static and dynamic objects with an autonomous floor scrubber. Intell. Serv. Robot. **13**(3), 403–417 (2020)
59. Yan, Z., Sun, L., Duckett, T., Bellotto, N.: Multisensor online transfer learning for 3d lidar-based human detection with a mobile robot. In: Proceedings of the 2018 IEEE/RSJ International Conference on Intelligent Robots and Systems (IROS), pp. 7635–7640. Madrid, Spain (2018)
60. Yan, Z., Sun, L., Krajnik, T., Duckett, T., Bellotto, N.: Towards long-term autonomy: a perspective from robot learning. In: Proceedings of the AAAI-23 Bridge Program on AI and Robotics, pp. 1–4. Washington, USA (2023)
61. Yan, Z., Sun, L., Krajnik, T., Ruichek, Y.: EU long-term dataset with multiple sensors for autonomous driving. In: Proceedings of the 2020 IEEE/RSJ International Conference on Intelligent Robots and Systems (IROS), pp. 10697–10704. Las Vegas, USA (2020)
62. Yang, B., Luo, W., Urtasun, R.: PIXOR: real-time 3d object detection from point clouds. In: IEEE Conference on Computer Vision and Pattern Recognition (CVPR), pp. 7652–7660 (2018). https://doi.org/10.1109/CVPR.2018.00798
63. Yang, R., Yan, Z., Yang, T., Ruichek, Y.: Efficient online transfer learning for 3d object classification in autonomous driving. In: Proceedings of the 2021 IEEE International Conference on Intelligent Transportation Systems (ITSC), pp. 2950–2957. Indianapolis, USA (2021)
64. Yang, R., Yan, Z., Yang, T., Wang, Y., Ruichek, Y.: Efficient online transfer learning for road participants detection in autonomous driving. IEEE Sens. J. **23**(19), 23522–23535 (2023)
65. Yang, R., Yang, T., Yan, Z., Krajnik, T., Ruichek, Y.: Preventing catastrophic forgetting in continuous online learning for autonomous driving. In: Proceedings of the 2024 IEEE/RSJ International Conference on Intelligent Robots and Systems (IROS), pp. 1–8. Abu Dhabi, UAE (2024)
66. Yang, T., Li, Y., Ruichek, Y., Yan, Z.: LaNoising: a data-driven approach for 903 nm ToF LiDAR performance modeling under fog. In: Proceedings of the 2020 IEEE/RSJ International Conference on Intelligent Robots and Systems (IROS), pp. 10084–10091. Las Vegas, USA (2020)
67. Yang, T., Li, Y., Ruichek, Y., Yan, Z.: Performance modeling a near-infrared ToF lidar under fog: a data-driven approach. IEEE Trans. Intell. Transp. Syst. **23**(8), 11227–11236 (2021)
68. Yang, T., Li, Y., Zhao, C., Yao, D., Chen, G., Sun, L., Krajnik, T., Yan, Z.: 3D ToF LiDAR in mobile robotics: a review. CoRR (2022). http://arxiv.org/abs/2202.11025
69. Yang, T., Yu, Q., Li, Y., Yan, Z.: Learn to model and filter point cloud noise for a near-infrared ToF lidar in adverse weather. IEEE Sens. J. **23**(17), 20412–20422 (2023)
70. Yao, D., Liu, B., Yang, R., Yan, Z., Fu, W., Yang, T.: Few-shot online learning for 3d object detection in autonomous driving. In: Proceedings of the 2023 International Conference on Autonomous Unmanned Systems (ICAUS). Nanjing, China (2023)
71. Zermas, D., Izzat, I., Papanikolopoulos, N.: Fast segmentation of 3d point clouds: a paradigm on lidar data for autonomous vehicle applications. In: IEEE International Conference on Robotics and Automation (ICRA), pp. 5067–5073 (2017). https://doi.org/10.1109/ICRA.2017.7989591

72. Zhao, Y., Zhang, X., Huang, X.: A technical survey and evaluation of traditional point cloud clustering methods for lidar panoptic segmentation. In: IEEE/CVF International Conference on Computer Vision Workshops (ICCV Workshops), pp. 2464–2473 (2021). https://doi.org/10.1109/ICCVW54120.2021.00279
73. Zhou, Y., Tuzel, O.: Voxelnet: end-to-end learning for point cloud based 3d object detection. In: IEEE Conference on Computer Vision and Pattern Recognition (CVPR), pp. 4490–4499 (2018). https://doi.org/10.1109/CVPR.2018.00472

Chapter 4
Robot Learning

Abstract This chapter introduces the research on robot learning, with a focus on Robot Online Learning (ROL) frameworks. It begins by defining ROL and motivating its necessity for robots operating in dynamic environments. Two prominent ROL frameworks are then presented: one based on Positive–Negative (P–N) learning and the other leveraging knowledge transfer. A comparative analysis highlights the strengths and weaknesses of each approach. Specifically, while P–N learning operates autonomously, it is susceptible to self-bias. Conversely, knowledge transfer can mitigate self-bias but requires external guidance and must address potential conflicts between internal and external knowledge sources. The chapter further explores strategies for mitigating catastrophic forgetting, a critical challenge in long-term ROL. Finally, it demonstrates how ROL can be applied to enhance socially-compliant robot navigation in extended, cross-environment deployments.

Keywords Online learning · Continual learning · Transfer learning · Catastrophic forgetting · Socially-compliant navigation

4.1 Introduction

A key factor in humanity's dominance on Earth is our remarkable capacity for learning. This ability allows us to accumulate knowledge, develop skills, and solve problems, enabling us to survive and thrive in a constantly evolving environment. Inspired by this, the development of robot intelligence incorporating **embodied learning** is a promising avenue of research, particularly given the significant role embodied experience plays in human intelligence. This approach offers several key advantages:

- It enables robots to learn through interaction with the physical world, facilitating the execution of complex real-world tasks.
- It fosters a deeper understanding of the world, crucial for long-term autonomy and online adaptability of robots.
- It promotes more effective learning by allowing robots to learn from their own experiences, rather than relying solely on human instruction or pre-existing data.

Embodied learning encompasses a broad spectrum, from the micro-level acquisition of specific knowledge or skills to the macro-level pursuit of lifelong learning, mirroring human capabilities.

This chapter focuses on embodied learning, specifically Robot Online Learning (ROL). Generally speaking, Online Learning (OL) is a machine learning paradigm where data is sequentially acquired and used to incrementally update a predictor for future data, contrasting with batch learning methods that train a predictor on the entire dataset at once. OL finds application in diverse areas, such as predicting user preferences for targeted advertising and product recommendations by internet companies. Within mobile robotics, the "online" aspect of OL emphasizes robot autonomy, which means that robots should learn in-situ, on-the-fly, spontaneously, and automatically, without human intervention. The remainder of this chapter provides a systematic overview of ROL.

4.2 Why Study Robot Online Learning

The motivation for studying ROL stems from two primary considerations. On the one hand, we believe that robots with autonomous learning capabilities are very consistent with people's imagination of future robots. Specifically, as a learning individual, the robot should show a certain degree of initiative in learning knowledge and applying learning outcomes. This need is especially evident when the deployment environment of the mobile robot is changing, or when certain tasks require the robot to travel across different environments. In such scenarios, the robot can sense the environmental changes through its perception system and learn about them online, ultimately responding appropriately to these changes in a timely manner. On the other hand, we are concerned about some limitations of offline learning methods. First, offline learning is usually accompanied by obvious human costs, such as data collection and annotation, model debugging and maintenance, etc. Second, offline models are inherently unable to support long-term autonomous operation of mobile robots, because there will always be situations that the robot has not seen or learned, such as corner cases, long-tail problems, domain shift, etc. For instance, Fig. 4.1 illustrates atypical road users, whose detection and tracking may pose a challenge for autonomous vehicles. Even a comprehensive Operational Design Domain (ODD) cannot anticipate every possible scenario.

4.3 Challenges of Robot Online Learning

This chapter addresses two key challenges in ROL.

1. **Autonomous sample extraction from sensor data:** Mobile robots rely on diverse sensors to perceive their internal state and surrounding environment.

Fig. 4.1 Some long-tail examples in road participant detection

The measurements from sensors are represented in various data forms, such as images produced by cameras and point clouds produced by 3D lidars. These data correspond to observations made by the robot and are analyzed by it to extract information of interest. Taking object detection introduced in Chap. 3 as an example, the robot needs to determine the location and category of the object it wants to learn in each observation, and then extract data representing the object as a learning sample. Automating this process is very challenging, particularly in complex and highly dynamic environments such as university cafeterias [1, 2] and urban roads [3, 4]. If we consider the 3D lidar introduced in Chap. 3, it is even more challenging due to the sparsity of the data it produces and the lack of easy-to-learn features such as color and texture.

2. **Mitigating catastrophic forgetting in long-term ROL:** In ROL, or even in any learning method that requires updating a model, new learning can cause the performance of previously learned models to degrade, a problem known as "catastrophic forgetting". A practical example is illustrated in Fig. 4.2. The probability of this problem occurring increases with the diversity of learning samples or tasks. In mobile robotics, the long-term deployment of robots will inevitably lead to the diversity of learning tasks. The problem of catastrophic forgetting has a long history of research in machine learning and is also one of the research focuses of the deep learning community in this era. However, despite significant progress in specific areas, many approaches to overcome catastrophic forgetting do not generalize directly to robotics or are simply not feasible due to limited onboard memory and computational resources. Therefore, there is a need to develop methods applicable to mobile robots, which is of interest in this book.

In addition, while the primary research objective is to address the challenges of robot social navigation, Sect. 4.6 also explores, within the context of current technological advancements, the integration of deep learning with ROL. A key challenge in the latter lies in efficiently updating computationally demanding deep neural network models—which offer the potential to break through the performance bottleneck—onboard robots with constrained computing resources, all while maintaining real-time performance.

For a better understanding, we further explain OL with reference to offline learning and incremental learning. The intuitive differences between the three can be seen

Fig. 4.2 Road participants of the same category in East Asia (left) and Western Europe (right) with very different appearances. After the robot learns in East Asian scenes and then updates the previously learned model in Western European scenes, the object detection performance of the model may deteriorate after returning to the East Asian scene, and vice versa

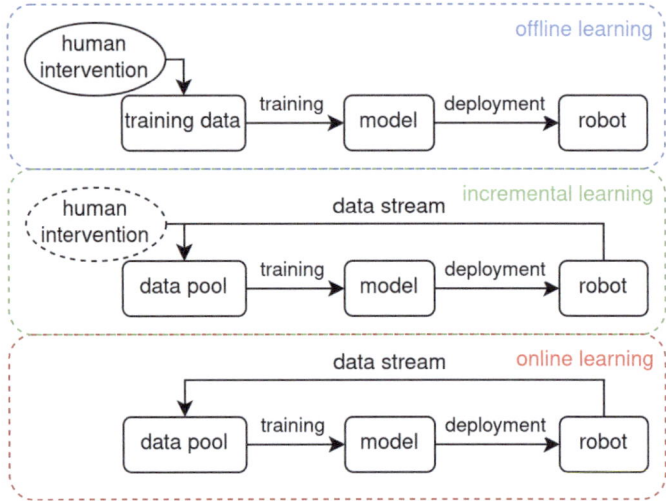

Fig. 4.3 The use of three different learning paradigms in mobile robotics

from Fig. 4.3. Offline learning is similar to offline programming commonly seen in industrial robotic arms, which means that the model is fully trained before being deployed to the robot and remains unchanged during the operation of the robot. To this end, data is collected in advance and usually annotated to ensure the final model performance. The typical workflow involves a sequence of steps: data collection, data annotation, model building, model training, model tuning, and model deployment. If the model needs to be updated, some or all of these steps will need to be repeated.

Incremental learning can be implemented both online and offline. It processes continuous data, but without strict real-time constraints. Human intervention is also permissible to guide the iterative learning process and ensure model performance. This learning paradigm prioritizes knowledge retention and mitigating catastrophic forgetting. In contrast, OL emphasizes autonomous, real-time learning during robot operation, without human intervention. Timeliness is a key characteristic of this learning method, which means learning quickly and applying the learned model immediately.

All object detection methods discussed in Sect. 3.3, with the exception of the two examples using Support Vector Machine (SVM) and Random Forest (RF), are non-online. To overcome the reliance on complete, annotated datasets and manual intervention, several approaches have been proposed. For instance, Shackleton et al. [5] employed surface matching for human detection, coupled with an Extended Kalman Filter (EKF) to predict the position of a human target and facilitate the detection in subsequent lidar scan. Teichman et al. [6] introduced a semi-supervised learning method for multi-object classification, requiring only a small set of manually labeled seed object tracks for classifier training. Dewan et al. [7] presented a classifier-free approach for dynamic object detection and tracking, relying on motion cues. Consequently, this method is not suitable for slow-moving or static objects, such as pedestrians.

The work presented in the following sections systematically reorganizes and expands upon research presented in [1, 3, 8–10]. Specifically, Yan et al. [1] introduced a ROL framework for human classification in 3D lidar scans, leveraging a multi-target tracker to eliminate the need for human experts to annotate the sensor data. Building upon this, Yan et al. [9] addressed limitations in [1] related to assumptions about human volume (as seen in Eq. 3.7) and movement speed. This was achieved by using one sensor (or model) to train another, thereby enhancing the generality and robustness of the ROL framework. Subsequently, Yang et al. [3] applied the framework from [9] to autonomous driving, achieving ROL for road participants, including cars, cyclists, and pedestrians. This work significantly enhanced the implementation from [1, 9] through the introduction of Online Random Forest (ORF) method. Further development in [10] resulted in a generalizable framework to mitigate catastrophic forgetting in long-term ROL. Most recently, Okunevich et al. [8] integrated deep learning models into the ROL framework for long-term, cross-environmental, socially-compliant robot navigation.

Research on catastrophic forgetting can be traced back to the 1990s [11, 12]. One approach to mitigating this issue involves preserving past knowledge by limiting changes to model weights. For instance, a memory buffer [13] can be employed to store data or gradient records from past training, thereby constraining the updates in the current learning process. In situations where retaining information from previous tasks is impractical due to privacy or resource constraints, regularization-based methods [14–16] offer an alternative by employing carefully designed regularization losses to constrain forgetting of previously acquired knowledge while learning new data. Another intuitive approach involves constructing a sufficiently large model and allocating a dedicated subset of the model to each task. This can be implemented by freezing a shared trunk and adding task-specific branches, effectively separating old and new knowledge. However, this strategy can lead to a rapid increase in model size [17]. Finally, replay-based approaches rely on retaining, or compressing, data representative of past tasks [18–20]. These methods combat forgetting by reintroducing stored samples during training on new tasks. These replayed samples play a critical role in joint training or loss optimization, safeguarding previously acquired knowledge.

Section 4.5 introduces Long Short-Term Online Learning (LSTOL) [10], an ensemble learning framework comprising a set of short-term learners and a long-term control mechanism. The former can be any model suitable for OL, such as capable of rapid iteration without storing learning samples. The latter features a gate controller that controls whether each existing short-term learner should be updated, retained, or deleted, or whether a new short-term learner should be created. The design of the controller is based on primitives rather than complex reasoning, fully considering the real-time requirements of robot physical interaction in dynamic real-world environments. Furthermore, unlike Long Short-Term Memory (LSTM) networks [21], LSTOL focuses on learning strategy rather than network architecture, and makes no assumptions about the temporal continuity of learning data and accommodates diverse short-term learner models.

Thanks to advances in deep learning, robot social navigation performance is continually improving [22–24]. Despite these advancements, deploying real-time online-updatable deep learning models on resource-constrained robot platforms remains a significant challenge. Deep learning models typically require substantial computational resources and memory, which can be prohibitive for embedded devices. While model compression and the use of more powerful edge devices are potential solutions, this book focuses on model optimization, specifically, designing deep learning architectures suitable for OL. To this end, Sect. 4.6 proposes a hierarchical structure combining a heavyweight network with a lightweight network to enable mobile robots to adapt autonomously and efficiently to new social environments. The heavyweight network provides basic, robust navigation control and remains static. The lightweight network evaluates the output of the heavyweight one and adjusts it for social compliance, when necessary. This network is updated online by analyzing, in real-time, the difference in social attributes between the robot's trajectory and the trajectories of surrounding humans, ultimately allows the robot to learn new social contexts when they differ from previously learned ones.

4.4 Autonomous Sample Generation

This section introduces two methods for autonomously generating learning samples, one is based on Positive–Negative (P–N) learning and the other is based on knowledge transfer.

4.4.1 P–N Learning

The first method [1, 2] consists of four modules: a cluster detector based on the adaptive clustering method introduced in Sect. 3.3.1.1, a multi-target tracker introduced in Sect. 3.4, a human classifier introduced in Sect. 3.3.2, and a sample generator based on P–N learning [25]. Specifically, as illustrated in Fig. 4.4, the point cloud data generated by the lidar is first input into the clustering module, which outputs candidate learning samples (i.e., clusters) to both the multi-target tracker and the online-updatable human classifier. The former is responsible for associating multiple samples belonging to the same target and estimating the instantaneous velocity of the target. The latter estimates the category to which each sample belongs, such as "human" or "non-human". After receiving the outputs of the multi-target tracker and the human classifier, the sample generator employs the P–N learning method to determine the final learning samples.

Because ROL operates without human supervision, it is inherently susceptible to errors, such as false positives and false negatives. While reinforcement learning could be considered for scenarios permitting trial-and-error with effective feedback, it falls outside the scope of the research presented here. Instead, we aim for the robot to autonomously detect and correct these errors. Drawing inspiration from the tracking-learning-detection paradigm [26], two independent "experts" are employed to correct the results given by the human classifier. The positive (P) expert corrects false negatives, converting them to positive samples, while the negative (N) expert corrects false positives, converting them to negative samples. Specifically, at each time step, the P-expert analyzes clusters currently classified as non-human, identifying likely false negatives and adding them to the training set as positive samples. Simultaneously, the N-expert examines clusters currently classified as human,

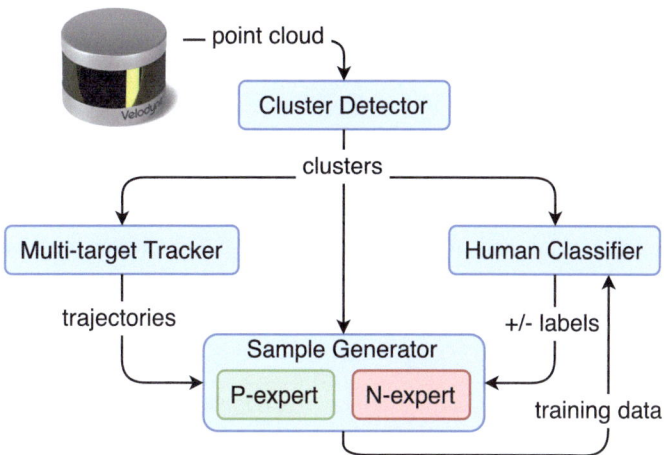

Fig. 4.4 P–N learning-based ROL framework

Algorithm 1 Iterative model training with P–N learning

Require: H: Human classifier,
    C: Clusters,
    C_h: Human clusters,
    C_n: Non-human clusters,
    S_p: Positive samples,
    S_n: Negative samples,
    T_p: Positive training set,
    T_n: Negative training set,
    p: Size of positive training set,
    n: Size of negative training set
Ensure: H: Human classifier
1: **while** True **do**
2:    **while** $|T_p| < p$ **or** $|T_n| < n$ **do**
3:       $[C_h, C_n] \leftarrow \text{classify}(H, C)$
4:       $S_p \leftarrow P_\text{expert}(C_n)$
5:       $S_n \leftarrow N_\text{expert}(C_h)$
6:       $T_p \leftarrow T_p \cup S_p$
7:       $T_n \leftarrow T_n \cup S_n$
8:    **end while**
9:    $\text{train}(H, T_p, T_n)$
10: **end while**

identifying likely false positives and adding them to the training set as negative samples. This dynamically constructed training set is then used to initialize and update the human classifier. Through iterative learning, the classifier's performance gradually improves. In practice, the P-expert enhances the classifier's generality, while the N-expert improves its discriminability. Implementation details are provided in Algorithm 1.

As for the implementation of the experts, it adheres to the structural constraints of P–N learning and employs rule-based (heuristic) reasoning. This approach offers advantages including efficiency, ease of understanding and maintenance, as well as strong interpretability and scalability. However, it also has limitations, notably limited generalizability and difficulty in handling incomplete, uncertain, and ambiguous information. Specifically, both the P-expert and N-expert rely on information from the multi-target tracker, making the tracker's performance crucial.

The P-expert operates according to the following rules:

1. Sustained tracking and minimum displacement: The cluster must be continuously tracked for a time interval of $K\Delta t$, during which it covers a minimum distance r_min^p:

$$r_k = \sqrt{(x_k - x_{k-1})^2 + (y_k - y_{k-1})^2} \quad \text{and} \quad \sum_{k=1}^{K} r_k \geq r_\text{min}^p \qquad (4.1)$$

2. Velocity constraints: The cluster's velocity must be non-zero and within a human's preferred walking speed ($v_\text{max}^p = 1.4$ m/s) [27]:

4.4 Autonomous Sample Generation

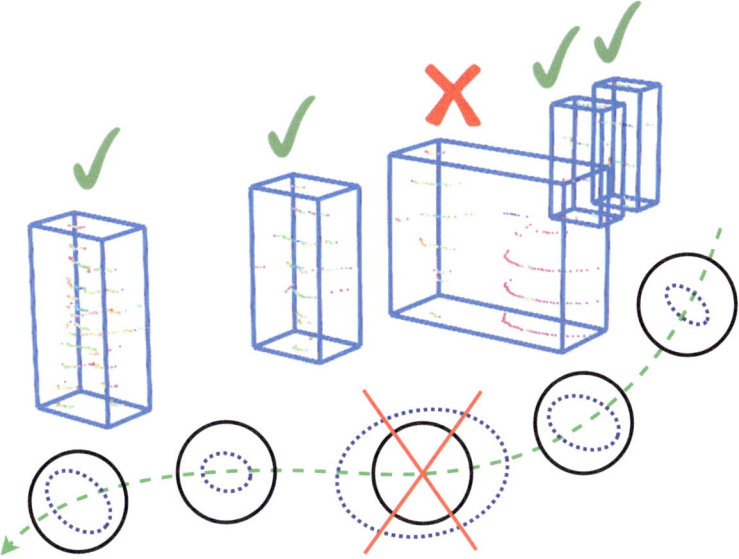

Fig. 4.5 Example of a human track sample containing mixed objects. The green dashed line represents the target's trajectory, and the blue dashed circle indicates the uncertainty in its position

$$v_k = \sqrt{\dot{x}_k^2 + \dot{y}_k^2} \quad \text{and} \quad v_{\min}^p \leq v_k \leq v_{\max}^p \quad (4.2)$$

3. Position variance constraint: The variances (σ_x^2, σ_y^2) of the cluster's estimated position (x_k, y_k) must satisfy:

$$\sigma_x^2 + \sigma_y^2 \leq (\sigma_{\max}^p)^2 \quad (4.3)$$

The values of K, $r_{\min}^p$, $v_{\min}^p$, and $\sigma_{\max}^p$ require empirical tuning prior to deployment. The position variance rule is particularly useful for distinguishing clusters where humans and other objects are merged (often due to under-segmentation, as shown in Fig. 4.5) or for identifying non-human clusters exhibiting sudden movements.

The N-expert operates according to the following rule:

1. Non-static condition: The cluster must not be completely static:

$$r_k \leq r_{\max}^n, \quad v_k \leq v_{\max}^n, \quad \text{and} \quad \sigma_x^2 + \sigma_y^2 \leq (\sigma_{\max}^n)^2 \quad (4.4)$$

The values of $r_{\max}^n$, $v_{\max}^n$, and $\sigma_{\max}^n$ require empirical tuning prior to deployment.

Following [26], a stability analysis of the ROL process can be performed by examining the variations in false positives (α) and false negatives (β) produced by the human classifier:

$$\begin{bmatrix} \alpha_{k+1} \\ \beta_{k+1} \end{bmatrix} = \begin{bmatrix} 1 - R^- & \frac{1-P^+}{P^+} R^+ \\ \frac{1-P^-}{P^-} R^- & 1 - R^+ \end{bmatrix} \times \begin{bmatrix} \alpha_k \\ \beta_k \end{bmatrix} \quad (4.5)$$

where k denotes the learning/training iteration, and P^+, R^+, P^-, and R^- represent the P-precision, P-recall, N-precision, and N-recall of the experts, respectively. The recursive equation in Eq. 4.5 constitutes a discrete dynamical system, expressible as $x_{k+1} = M x_k$. It can be demonstrated that if both eigenvalues (λ_1 and λ_2) of M are less than one, then x converges to zero.

An interesting observation by looking at Eq. 4.5 is that, if $P^+ = P^- = 1$, then M becomes a diagonal matrix, and therefore its eigenvalues are simply $\lambda_1 = 1 - R^-$ and $\lambda_2 = 1 - R^+$. In this case, even a small recall for both experts is sufficient to guarantee $\lambda_1 < 1$ and $\lambda_2 < 1$, thus ensuring the stability of the learning process. Based on these criteria, the following performance metrics are used for the P–N experts:

$$\text{P-expert Precision:} \quad P^+ = \frac{n_t^+}{n_t^+ + n_f^+} \quad (4.6)$$

$$\text{P-expert Recall:} \quad R^+ = \frac{n_t^+}{\beta} \quad (4.7)$$

$$\text{N-expert Precision:} \quad P^- = \frac{n_t^-}{n_t^- + n_f^-} \quad (4.8)$$

$$\text{N-expert Recall:} \quad R^- = \frac{n_t^-}{\alpha} \quad (4.9)$$

where n_t^+ and n_f^+ represent the number of true positives and false positives, respectively, identified by the P-expert, while n_t^- and n_f^- represent those identified by the N-expert. α represents the total number of false positives from the human classifier, which the N-expert should correct. Similarly, β represents the total number of false negatives from the human classifier, which the P-expert should correct.

4.4.2 Knowledge Transfer

As previously mentioned, the performance of P–N learning is highly dependent on the structural constraints of the data. However, the dynamic environments in which mobile robots operate are sometimes weakly structured or may even lack discernible structure. The advancements in robot multimodal perception, particularly the breakthrough performance achieved with cameras, have inspired a solution to

4.4 Autonomous Sample Generation

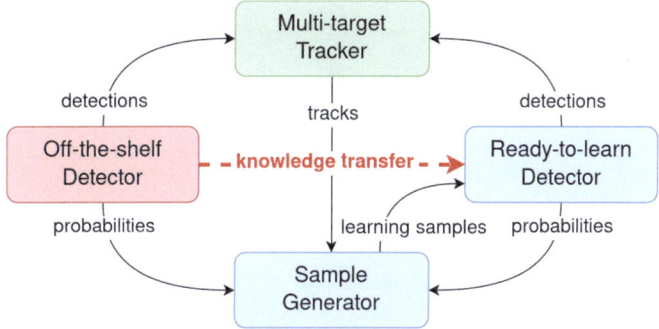

Fig. 4.6 Knowledge transfer-based ROL framework

this challenge: **using easily-trained sensors to train those that are more difficult**. Consequently, a second sample generation method is proposed [9], also comprising four modules (as illustrated in Fig. 4.6): an off-the-shelf detector D_s, a ready-to-learn detector (for OL) D_d, a multi-target tracker T, and a sample generator G.

This approach can be further formalized as a transfer learning process [28]: Given a source domain $D_s = \{X_s, M_s(X)\}$ and a learning task L_s, and a target domain $D_d = \{X_d, M_d(X)\}$ and a learning task L_d, transfer learning leverages knowledge from D_s and L_s to facilitate learning the target predictive model $M_d(\cdot)$ in D_d, where $D_s \neq D_d$ or $L_s \neq L_d$. In the knowledge transfer-based ROL framework, the source learning task L_s is performed in advance, whether online or offline, while the target learning task L_d is performed online with the assistance of $M_s(X)$.

Specifically, D_s and D_d output detections to T and probabilities corresponding to the detections to G. D_s contributes to multi-target tracking on the one hand and helps D_d learn on the other. D_d also helps with tracking, which may be relatively small at first, but should increase as learning iterates. From a machine learning perspective, D_s provides labeled data, while D_d provides both labeled and unlabeled data. T associates the received detections to generate a track V^k. Both moving and non-moving targets are tracked. For the latter, the track length is a very small value close to zero. The tracker plays a crucial role in knowledge transfer within this ROL framework because it correlates detections, effectively establishing links between the different detectors, thus enabling D_d to learn from D_s.

The sample generator G fuses information from D_s, D_d, and T to generate learning samples for D_d. This constitutes a form of multi-sensor post-fusion, where resolving information conflicts presents a significant challenge. For instance, a dummy model might be accurately identified by the camera but misclassified as a human by the lidar. To address this, and drawing inspiration from multi-sensor occupancy grid map fusion in Simultaneous Localization and Mapping (SLAM) [29, 30], a *track probability* estimation method (visualized in Fig. 4.7) is proposed based on Bayes' theorem. This method measures the likelihood that a track belongs to a specific object category, such as "human", and is defined as follows. Consider a

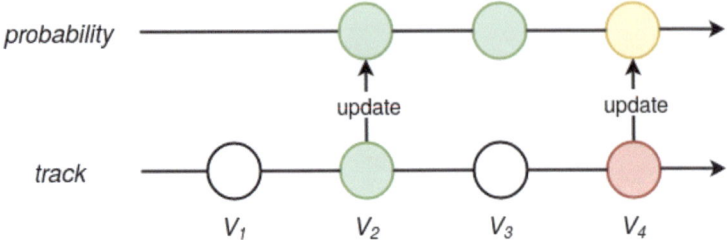

Fig. 4.7 Schematic diagram of the track probability

tracklet $V^k = \{(x_1, y_1), \ldots, (x_k, y_k)\}$ of length k generated by the tracker T. Let $P(Y^k|V^k)$, simplified to $P(Y)$, denote the probability that this tracklet belongs to a certain object category Y. Let $d_i^j \in D$ (where $D = D_s \cup D_d$) represent the detections produced by detector j at time i, which comprise the tracklet V^k. Given o detectors, we aim to compute the probability of V^k belonging to category Y, i.e., $P(Y|\{d_i^j\})$ for $i = 1, \ldots, k$ and $j = 1, \ldots, o$. Assuming Markov independence, such that $P(d_i^j|Y, \{d_m^n\}) = P(d_i^j|Y)$ for $m = 1, \ldots, i-1, i+1, \ldots, k$ and $n = 1, \ldots, j-1, j+1, \ldots, o$, and starting with an initial probability $P(Y) = 0.5$, we can calculate $P(Y|\{d_i^j\})$ as follows:

$$P(Y|\{d_i^j\}) = \frac{\text{odds}(Y|\{d_i^j\})}{1 + \text{odds}(Y|\{d_i^j\})} \quad (4.10)$$

where

$$\text{odds}(Y|\{d_i^j\}) = \prod_{j=1}^{o}\prod_{i=1}^{k} \text{odds}(Y|d_i^j) \quad (4.11)$$

and

$$\text{odds}(Y|d_i^j) = \frac{P(Y|d_i^j)}{P(\neg Y|d_i^j)} = \frac{P(Y|d_i^j)}{1 - P(Y|d_i^j)} \quad (4.12)$$

This probabilistic fusion approach has been theoretically and practically shown to account for any existing knowledge about sensor interactions [29, 31], making it particularly well-suited for track category estimation based on multi-sensor information fusion.

An intuitive and effective method for labeling the track V^k is to threshold $P(Y|\{d_i^j\})$, i.e., $P(Y^k|V^k, D)$. Specifically, given a predefined threshold P^*, the label for each detection $(x_i, y_i) \in V^k$ is determined as follows:

$$y_i = \begin{cases} z \in \mathbb{Z}^+, & \text{if } P(Y|\{d_i^j\}) \geq P^* \\ z \in \mathbb{Z}^-, & \text{if } P(Y|\{d_i^j\}) < 1 - P^* \end{cases} \quad (4.13)$$

4.4 Autonomous Sample Generation

where $\mathbb{Z}^+$ and $\mathbb{Z}^-$ represent the sets of positive and negative integers, respectively. It is crucial to note that only P^* is used to determine whether all detections within a track V^k are classified as positive or negative (due to $P(Y = 0|\cdot) = 1 - P(Y = 1|\cdot)$). Using any other threshold or an "otherwise" assignment would introduce nonlinearities and complex special cases.

Once a track is labeled, learnable samples become available. Batch-incremental Training (BiT) [32] strategy is proposed for model learning in D_d for two reasons: *a*) $|V^k| \geq 1$, and *b*) it offers greater model flexibility and accommodates concept drift in ROL. Formally, given a training set sequence $X_1^k, X_2^k, \ldots, X_m^k$, where

$$X_i^k = \{(x_i, y_i) | x_i \in U_i \subseteq \mathbb{R}^n, y_i \in Y_i \subseteq \mathbb{Z}\}, \quad 1 \leq i \leq m \quad (4.14)$$

where Y_i represents the set of category labels in the training set X_i (e.g., human or non-human), and k denotes the batch window size. Let M_1 represent the model trained on X_1. The BiT process is then defined as:

$$\text{BiT}(X_i, M_{i-1}) = M_i, \quad 2 \leq i \leq m \quad (4.15)$$

where the subscript i represents the number of iterations rather than time.

It can be seen that the knowledge transfer-based framework exhibits the characteristics of semi-supervised learning, since (1) the model sequentially learns non-independently and identically distributed (non-IID) $P(V^k)$ fully labeled by G, and (2) the model learns from both labeled and unlabeled data in a batch manner. It can also be seen that the framework is inclusive, allowing the integration of various learning models such as Support Vector Machine (SVM), Random Forest (RF), Deep Neural Network (DNN), and so forth.

Unlike the stability analysis for P–N learning, analyzing the stability of knowledge transfer-based ROL at a micro-level (i.e., a single module) is challenging. Therefore, stability is analyzed at a macro-level (i.e., the entire framework). Let u_i denote the number of correctly predicted labels by G in the i-th learning iteration. Learning stability is then defined as:

$$\text{stability}(I) = \sum_{i=1}^{I} \|u_i - u_{i+1}\| \quad (4.16)$$

Theoretically, according to Lyapunov stability, the iterative system stabilizes if:

$$\lim_{I \to \infty} \frac{\text{stability}(I)}{I} = 0 \quad (4.17)$$

4.5 Mitigation of Catastrophic Forgetting

In long-term ROL, catastrophic forgetting typically occurs in two scenarios: an increase in the number of classes or tasks to be learned, and a shift in knowledge domains. This section focuses on addressing the challenges posed by the latter. To this end, the LSTOL framework [10] is presented, which aims to adapt a model to new data distributions without forgetting the knowledge learned from previous domains. Specifically, as illustrated in Fig. 4.8, LSTOL comprises a set of short-term learners and a long-term controller. Each short-term learner can be implemented as a model such as SVM, RF, DNN, etc., and learns from streaming data of various modalities, such as images and point clouds. The long-term controller supervises the learning process of the short-term learners and performs three key functions:

- *Information Collection:* This function gathers information about the short-term learners, including their current output confidence, accuracy, and activity level on downstream tasks. This information forms the basis for decisions regarding knowledge retention and new knowledge acquisition.
- *Gating Control:* Based on an evaluation of the collected information and the predicted probabilities provided by the short-term learners, this function determines the appropriate actions for the framework. These actions include retaining, updating, or deleting existing learners, or creating new ones.
- *Weight Estimation:* This function dynamically adjusts the weights of each short-term learner based on its past performance. A learner with high accuracy on a task will have its influence (i.e., weight) on that task increased. Learners exhibiting high prediction confidence act as "experts" and primarily determine the final prediction. The task with the highest confidence becomes the output of both the long-term controller and the entire framework.

It is important to emphasize that the LSTOL framework operates in a learn-as-you-go fashion. The output of the long-term controller can be directly used for downstream tasks, such as the object detection task introduced in Sect. 3.3. A specific implementation of LSTOL, applied to road participant detection in point clouds as a downstream task, is shown in Fig. 4.9. The details are as follows.

4.5.1 Learning Sample Extraction

The learning samples are extracted from the point cloud generated by the 3D lidar installed on the self-driving car and are defined as follows:

$$S = \text{track}(\{x_i, c_i, t_i\}_{i=1}^{n} \bar{c}) \tag{4.18}$$

where $\{x_i, c_i, t_i\}_{i=1}^{n}$ represents a set of n instances (in the form of clusters) of object x tracked at different times t_i. c_i represents the confidence that instance i belongs

4.5 Mitigation of Catastrophic Forgetting

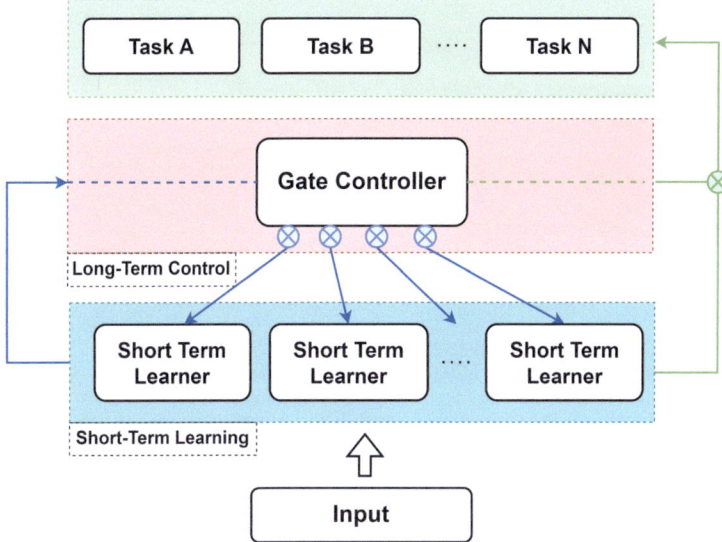

Fig. 4.8 Schematic diagram of the LSTOL framework. It consists of two modules: short-term learning and long-term control. The input sample is first predicted by a set of short-term learners. The long-term control module then collects these pre-prediction information and calculates quantitative indicators for online prediction. These indicators are subsequently input to the "Gate Controller", which determines the appropriate action for each learner (e.g., retain, update, remove, or create). During the learning phase (indicated by the blue line), the long-term control module calculates the online loss of each learner based on the current input sample to update the learner's weights. These updated weights are then used to determine the object category during the prediction phase (indicated by the green line)

to a certain object class, and $\bar{c}$ represents the overall confidence that the entire track belongs to that class. Specifically (as shown in Fig. 4.10), given an object track consisting of temporal detection samples from different detectors, the object can be detected by one or more detectors (such as point cloud-based and image-based), with corresponding confidence levels indicating the likelihood of the object belonging to a particular class. In principle, all learning samples within a track should belong to the same class, and this class label is determined by fusing the confidence scores of all samples in the track. For example, although the point cloud-based detector might occasionally misclassify an object (e.g., misclassifying a car as a cyclist due to the need to improve performance in the early stages of ROL), if the image-based detector and the majority of correct classifications from the point cloud-based detector indicate that the object is a car, the entire track will be labeled as a car sample. This car label will then be used for training the point cloud-based detector.

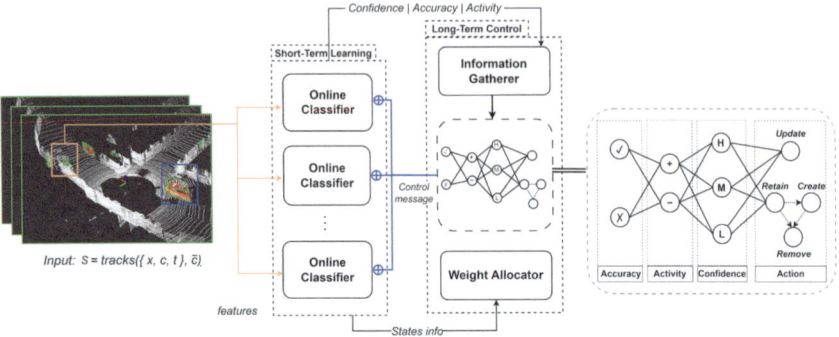

Fig. 4.9 Implementation overview of the LSTOL framework. From left to right: Object samples in successive point cloud frames are first associated using a multi-target tracker [4] and then input to the online classifiers within the short-term learning module. The classification results from these classifiers are collected by the *Information Gatherer* in the long-term control module, which evaluates the performance of each learner based on a matrix of confidence, accuracy, and activity. This evaluation is then input to the *Dynamic Gate Controller*, which determines the appropriate action for each classifier (e.g., retain, update, remove, or create). The classification loss for each sample is also input to the *Weight Allocator*, which updates the weights of the learners for different object categories. These weights are subsequently used to determine the final classification of each object

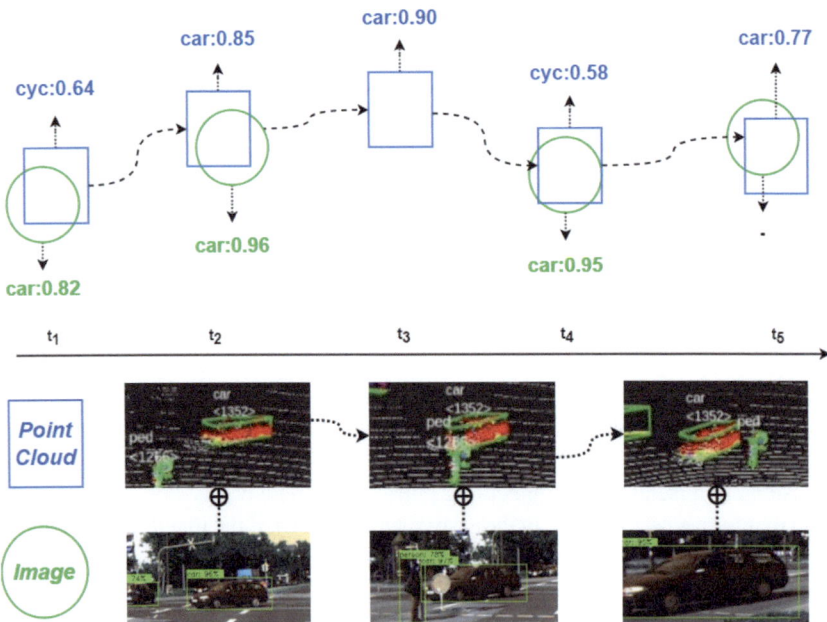

Fig. 4.10 Illustration of the extraction of learning samples. Rectangles represent detections by the point cloud-based detector that require OL, while circles denote detections by the off-the-shelf image-based detector

4.5.2 Short-Term Learning

In the short-term learning module (denoted as *stl*), each learner employs ORF [33] (detailed in Sect 3.3.2.2), which enables rapid training of multi-class models and their real-time deployment. This can be formalized as:

$$stl(x) = h \sum_{i=1}^{I} w_i \text{ORF}_i(x) \tag{4.19}$$

where w_i represents the weight of learner i, and h represents the fusion strategy determined by the long-term control module.

4.5.3 Long-Term Control

Information Gatherer

In ROL, due to the absence of ground truth, traditional metrics like precision and recall are impractical for real-time performance evaluation of the model. Therefore, three "online" metrics including confidence, accuracy, and activity, are employed to assess the real-time performance of each short-term learner. These are defined as follows:

$$\text{Confidence}_i = \max_{j \in \{1,\dots,J\}} (p_i^j) \tag{4.20}$$

where p_i^j denotes the predicted probability that learner i classifies an object as belonging to class j, and Confidence$_i$ represents the highest probability assigned by learner i across all classes.

$$\text{Accuracy}_i = \frac{p_i^{\text{correct}}}{p_i^{\text{total}}} \tag{4.21}$$

where p_i^{correct} represents the number of correct predictions made by learner i, and p_i^{total} represents the total number of predictions made by learner i.

$$\text{Activity}_i = \sum_{t=1}^{T} \text{update}_i(t) \tag{4.22}$$

represents the number of times learner i is updated within the time window T.

Dynamic Gate Controller

This module implements a probabilistic decision-making process to determine appropriate actions based on the three metrics provided by the Information Gatherer (as illustrated on the far right of Fig. 4.9), summarized in Algorithm 2. Specifically, line 3 stipulates that a learner must achieve high accuracy before updating to prevent learning from incorrect data. The "retain" operation (line 8) encompasses two scenarios. The first is when the learner exhibits low accuracy, indicating that the new sample's prediction falls outside its current knowledge domain. The second is when the learner demonstrates both high accuracy and high confidence, suggesting that it is already proficient with the input data and further learning is unnecessary. Line 16 aims to eliminate learners exhibiting low confidence, low accuracy, and low activity. However, as learner removal is a potentially detrimental operation, it is only performed when the number of learners reaches the predefined maximum and a new learner needs to be instantiated. Furthermore, while removal inevitably leads to some loss of previously acquired knowledge, a balance must be struck between this loss and allowing an unbounded number of learners. Line 26 specifies that if none of the existing learners have been updated, a new learner will be created.

Weight Allocator

For each learner, a Dynamic Expert Weights (DEW) table is constructed, assigning a weight to each class to represent the learner's classification ability for that class. Classes with better performance receive higher weights, thus amplifying their influence on the system's overall prediction. Specifically, given a new set of samples S input at time step $t + 1$, where each $s \in S$ is confidently assigned to class $k \in K$, the Kronecker delta, denoted as y_s, measures the consistency between the sample's predicted class and its true class k. Let $p_{s,k}$ denote the probability that learner i predicts sample s belongs to class k. The loss for sample s is then calculated using the logarithmic loss function:

$$L_{s,k} = -[y_s \log(p_{s,k}) + (1 - y_s) \log(1 - p_{s,k})] \quad (4.23)$$

The current weight $w_k(t)$ is updated using Exponentially Weighted Moving Average (EWMA) to reward correct predictions and penalize incorrect ones:

$$w_k(t + 1) = \lambda w_k(t) + (1 - \lambda) L_{\text{total}} \quad (4.24)$$

where

$$L_{\text{total}} = -\frac{1}{|S|} \sum_{s \in S} L_{s,k} \quad (4.25)$$

where $L_{s,k}$ represents the loss of sample s belonging to class k, and $|S|$ denotes the number of samples in S. The parameter λ controls the update speed of the weights,

4.5 Mitigation of Catastrophic Forgetting

Algorithm 2 Dynamic gate control

Require: N: Maximum number of learners,
 c_i: Confidence of learner i,
 a_i: Accuracy of learner i,
 v_i: Activity of learner i
Ensure: Learner i,
 Learner j,
 A new learner
1: $updated \leftarrow$ false
2: **for** each learner $i \in \mathcal{I}$ **do**
3: $\quad p \leftarrow \text{odds}(1 - c_i, a_i, 1 - v_i)$
4: $\quad$ **if** $p > 0.5$ **then**
5: $\quad\quad$ update(i)
6: $\quad\quad updated \leftarrow$ true
7: $\quad$ **else**
8: $\quad\quad$ retain(i) {No action taken}
9: $\quad$ **end if**
10: **end for**
11: **if** $\neg updated$ **then**
12: $\quad$ **if** $|\mathcal{I}| = N$ **then**
13: $\quad\quad p_{\max} \leftarrow 0$
14: $\quad\quad j \leftarrow \emptyset$
15: $\quad\quad$ **for** each learner $i \in \mathcal{I}$ **do**
16: $\quad\quad\quad p \leftarrow \text{odds}(1 - c_i, 1 - a_i, 1 - v_i)$
17: $\quad\quad\quad$ **if** $p > 0.5$ and $p > p_{\max}$ **then**
18: $\quad\quad\quad\quad j \leftarrow i$
19: $\quad\quad\quad\quad p_{\max} \leftarrow p$
20: $\quad\quad\quad$ **end if**
21: $\quad\quad$ **end for**
22: $\quad\quad$ **if** $j \neq \emptyset$ **then**
23: $\quad\quad\quad$ remove(j)
24: $\quad\quad$ **end if**
25: $\quad$ **else**
26: $\quad\quad$ create(new learner)
27: $\quad\quad \mathcal{I} \leftarrow \mathcal{I} \cup \{\text{new learner}\}$
28: $\quad$ **end if**
29: **end if**

---

balancing the learner's past and present accuracy. The final prediction stage employs a voting strategy called Hand-raised as Expert (HraE):

$$p_k = \sum_{i: p_{s,k} > \theta_c} w_{i,k} p_{s,k} \qquad (4.26)$$

where $w_{i,k}$ is the weight of learner i for class k, and θ_c is the minimum confidence threshold for a learner's prediction to be considered. This strategy prioritizes predictions from learners with higher weights, effectively allowing the long-term control module to focus on the most influential learners.

4.6 Robot Online Learning for Navigation

In this section, we demonstrate the versatility and efficacy of ROL by showcasing its application to a challenging downstream task: socially-compliant robot navigation. Building upon the object detection example presented in Sect. 3.3, we delve into how ROL empowers robots to navigate human-populated environments with appropriate social awareness. Socially-compliant robot navigation presents a unique set of challenges. Robots must seamlessly integrate into human spaces, adhering to often implicit social norms and adapting to diverse human behaviors. These behaviors are influenced by a multitude of factors, including individual preferences, cultural contexts, and dynamic environmental conditions. Predicting and exhaustively enumerating all possible scenarios for offline training is inherently difficult, if not impossible. Consequently, traditional offline learning-based approaches often struggle to guarantee socially acceptable robot behavior in long-term deployments, particularly when faced with novel or cross-environment situations. These methods may exhibit brittle performance when encountering situations not explicitly represented in the training data, leading to navigation that feels unnatural, unpredictable, or even intrusive to humans.

ROL offers a promising solution to this challenge by enabling robots to learn and adapt online, in real-time, to the nuances of human social dynamics [8]. This section introduces a two-layered ROL-based approach designed to imbue robots with social intelligence during navigation. This architecture mirrors the knowledge transfer-based ROL framework discussed in Sect. 4.4.2, leveraging a pre-trained "off-the-shelf" navigator as a foundation upon which a "ready-to-learn" social navigation layer is built. Specifically, the bottom layer employs a deep reinforcement learning (DRL) approach to provide the robot with fundamental navigation capabilities, generating basic movement commands. This DRL-based navigator can be pre-trained in simulation or on a limited real-world dataset, providing a robust foundation. However, it may lack the social finesse required for seamless human-robot interaction.

The upper layer, powered by ROL, addresses this limitation by refining the raw navigation commands from the DRL layer, injecting social awareness and ensuring compliance with human expectations. This ROL layer acts as a social filter, adapting the robot's trajectory and behavior based on observed human interactions and environmental cues. By continuously learning and updating its internal model of social norms online, the robot becomes increasingly adept at navigating in a socially appropriate manner, even in previously unseen environments. This online adaptation allows the robot to personalize its navigation style to different social contexts and individual preferences, fostering a more natural and comfortable interaction with humans. The overall architecture of this two-layered social navigation framework is visualized in Fig. 4.11. The following subsections will detail the specific implementations of both the DRL-based bottom layer and the ROL-based upper layer.

4.6 Robot Online Learning for Navigation

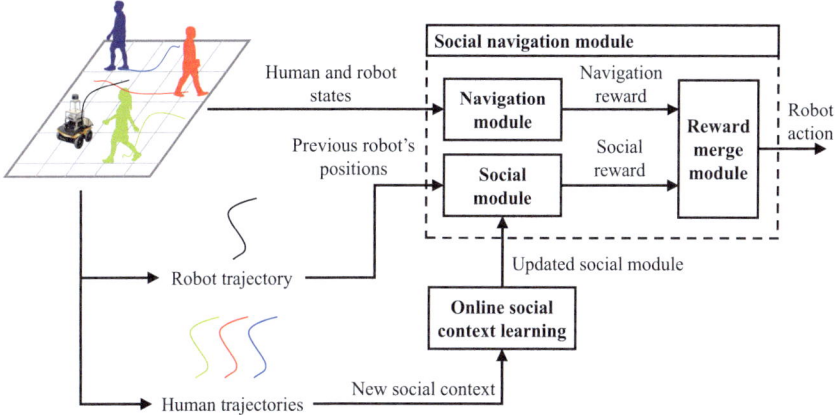

Fig. 4.11 Method design for enhancing the social navigation ability of robots using ROL. The navigation module is the bottom layer, which computes the next actions for the robot. The social module is the upper layer, which refines these actions to improve social compliance when needed. Online social context learning enables timely updates to the social module, facilitating socially normative robot navigation, particularly during long-term or cross-environment deployments

4.6.1 Basic Navigation Module

The bottom layer of the social navigation framework is implemented using Socially Attentive Reinforcement Learning (SARL) [22], a value-based deep reinforcement learning approach known for its strong performance in navigating human-populated environments. SARL aims to learn an optimal navigation policy, denoted as $\pi^*(s_t^{jn})$, that maximizes the cumulative discounted reward, effectively achieving the highest value $V^*(s_t^{jn})$ for the joint state s_t^{jn} at time t:

$$V^*(s_t^{jn}) = \sum_{i=t}^{T} \gamma^{i v_{\text{pref}}} R_t(s_t^{jn}, \pi^*(s_t^{jn})) \tag{4.27}$$

where $\gamma \in [0, 1)$ represents the discount factor, v_{pref} serves as a normalization parameter for γ, improving numerical stability during training [34]. and T denotes the time step of the final state. To approximate the next joint state $s_{t+\Delta t}^{jn}$, a constant velocity model (see Sect. 3.4.2) is employed to predict human movements over a short temporal interval Δt. This allows us to estimate the future state based on the current joint state s_t^{jn} and the robot's action a_t. Consequently, the optimal policy can be expressed as:

$$\pi^*(s_t^{jn}) = \arg\max_{a_t}[R_t(s_t^{jn}, a_t) + \gamma^{\Delta t v_{\text{pref}}} V(s_{t+\Delta t}^{jn})] \tag{4.28}$$

where the value network V is trained using the temporal difference (TD) method with experience replay [22–24]. The robot's action space is defined by its speed and angular direction. Holonomic kinematics are employed to model robot motion, allowing movement in any direction with variable acceleration. Specifically, the action space comprises 80 discrete actions, combining five linearly spaced velocities between 0 and v_{pref} with 16 evenly distributed angular directions from 0 to 2π. The best action is selected by evaluating the sum of the reward and value functions for each action in the action space.

A crucial modification introduced here to the original SARL framework lies in the design of the reward function. Instead of simply rewarding the robot for reaching the target position, we incentivize it to minimize unnecessary path deviations while maintaining social compliance. The original binary reward is therefore replaced with $d_{\text{plan}}/d_{\text{real}}$, where d_{plan} represents the initial Euclidean distance from the robot's starting position to the target, and d_{real} denotes the actual distance traveled by the robot upon reaching the target. This ratio encourages the robot to find a direct, efficient path. The complete reward function is defined as follows:

$$R_t(s_t^{jn}, a_t) = \begin{cases} -0.25 & \text{if } d_{\min} < 0 \text{ (collision)} \\ d_{\text{plan}}/d_{\text{real}} & \text{else if } d_g = 0 \text{ (goal reached)} \\ -0.1 + d_{\min}/2 & \text{else if } d_{\min} < d_c \text{ (close to human)} \\ 0 & \text{otherwise} \end{cases} \quad (4.29)$$

where $d_{\min}$ is the distance to the nearest person, and d_c represents the comfortable social distance. Critically, d_c can be dynamically adjusted based on the specific spatio-temporal context. For instance, d_c might be larger in crowded malls or during periods of heightened social distancing (e.g., during an epidemic), and smaller in less crowded environments or during normal times. This adaptability allows the robot to personalize its navigation behavior to different social situations. $d_g = 0$ indicates that the robot has reached the goal.

4.6.2 Online Adaptation Module

The upper layer of the framework comprises a ROL-based social adaptation module, which is designed to learn the nuances of social context from observed human trajectory data and subsequently refine the navigation control commands generated by the bottom layer (SARL). Crucially, to ensure adaptability to diverse social environments, this module is updated online, continuously incorporating new social context information. The core principle behind this online adaptation is intuitive: if the action a_t suggested by SARL is deemed socially acceptable, its corresponding value should be increased, thereby reinforcing its selection by the optimal policy $\pi^*(s_t^{jn})$. Conversely, if the action is considered socially inappropriate, its value should be decreased, discouraging its future selection. Building upon the effectiveness of small-batch ROL [2] and leveraging the power of tracklet-based behavioral analysis [35],

4.6 Robot Online Learning for Navigation

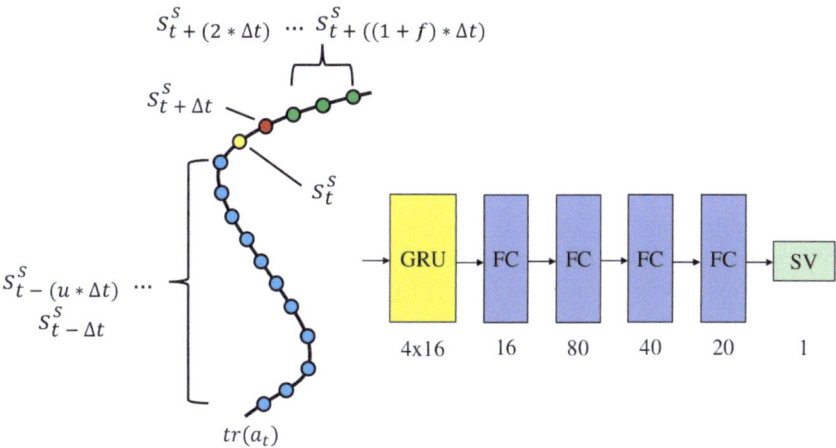

Fig. 4.12 A robot tracklet and the social value network. The blue dots represent u past robot states, the yellow dot indicates its current state, the red dot shows the robot's state after action a_t, and the green dots depict f predicted future states. The social value network, comprising a GRU and four fully connected (FC) layers, takes the robot tracklet as input. The numbers below indicate input dimensions. The network output, Social Value (SV), is 1 for a social tracklet and 0 for a non-social tracklet

a tracklet-based sociality assessment method is employed. This approach is visualized in Fig. 4.12. Specifically, a tracklet is defined as a sequence of the robot's instantaneous states:

$$tr_t(s_t, a_t) = [s^s_{t-(u\Delta t)}, s^s_{t-((u-1)\Delta t)}, \ldots, s^s_t, s^s_{t+\Delta t}, \ldots, s^s_{t+((1+f)\Delta t)}] \quad (4.30)$$

where each state s^s_t within the tracklet encapsulates the robot's position and velocity at time t, represented as $s^s_t = [p_x, p_y, v_x, v_y]_t$. The tracklet extends both backward and forward in time. The terms $s^s_{t-(u\Delta t)}, s^s_{t-((u-1)\Delta t)}, \ldots$ represent the robot's states at u previous time steps, while s^s_t denotes the current state. $s^s_{t+\Delta t}$ represents the robot's state immediately after executing action a_t at time t. Finally, $\ldots, s^s_{t+((1+f)\Delta t)}$ represents the predicted future states of the robot, assuming action a_t is repeatedly applied for f subsequent time steps. This forward-looking component allows the module to anticipate the short-term consequences of the chosen action.

A binary classification approach is employed to assess the social acceptability of each tracklet, categorizing them as either "social" or "non-social". This classification is achieved using a social value function, defined as:

$$SV_t(tr_t(s_t, a_t)) = \psi_{sv}(tr_t(s_t, a_t), W_{sv}) \quad (4.31)$$

where $\psi_{sv}(\cdot)$ represents a Gated Recurrent Unit (GRU) [36] combined with a Multi-Layer Perceptron (MLP), and W_{sv} denotes the weights of this combined model. As illustrated in Fig. 4.12, the GRU-MLP model consists of four fully connected layers, employing ReLU nonlinear activation functions and batch normalization. By incorporating the social value function into the action selection process, the final optimal strategy for robot navigation, augmented with the social module, becomes:

$$\pi^*(s_t^{jn}) = \arg\max_{a_t}[R_t(s_t^{jn}, a_t) + \gamma^{\Delta t v_{\text{pref}}} V(s_{t+\Delta t}^{jn}) + k_s SV_t(tr_t(s_t^{jn}, a_t))] \quad (4.32)$$

where k_s represents the weight or coefficient assigned to the social value during robot navigation. This coefficient allows for tuning the relative importance of social considerations compared to other factors, such as reward and value. The value of k_s can be adjusted based on various robot parameters, including speed, time step, and the specific social context. This tunability allows for fine-grained control over the robot's social behavior.

The online update mechanism for the social module is detailed in Algorithm 3. Initially, a buffer T_t is employed to continuously collect the states of both the robot and the surrounding humans. Once this buffer is filled (line 1), the states forming the robot's most recent tracklet and the states comprising the humans' most recent tracklets are processed separately (lines 2–3). When the number of stored robot tracklets reaches a predefined threshold (line 4), the effectiveness of the social module is evaluated.

A crucial step involves determining the labels (i.e., social or non-social) for the states within a subset of the most recent K_{up} human tracklets (lines 3 and 7). Traditional methods often rely on the simplifying assumption that all human behavior is inherently social [37–39]. Here, a more nuanced labeling approach is proposed. Specifically, the "extra distance ratio" metric (defined in Sect. 2.2.2.3) is used:

$$R_{\text{dist}} = \frac{d_s}{d_a} \quad (4.33)$$

where d_s represents the Euclidean distance between the start and end points of a tracklet, and d_a denotes the actual path length of the tracklet. According to [40], a higher R_{dist} value indicates more socially efficient movement. Consequently, if a tracklet's R_{dist} exceeds a predetermined threshold, all states within that tracklet are labeled as social, otherwise, they are considered non-social (line 5). This method provides the robot with information about the external social context. The robot then infers its internal social context through the output of the social module (line 6).

Finally, if the binary classification accuracy between the externally observed social context (from human tracklets) and the internally perceived social context (from the robot's social module) falls below a certain threshold (indicating a significant discrepancy, line 8), the social model is updated (line 12). The same R_{dist}-based labeling method is applied to label both the human tracklet set Tr_h (line 9) and the robot tracklet set Tr_r (line 10). The new training dataset D_{new} is then constructed, comprising the labeled human tracklets Tr_h and the non-social robot tracklets from Tr_r (line 11). Importantly, to mitigate overfitting, both the robot and human tracklet

Algorithm 3 Online social context learning

Require: $\psi_{sv}(\cdot, W_{sv})$: Social neural network,
T_t: Buffered human and robot states at time t,
L_{trak}: Input dimension of ψ_{sv},
Tr_r: Set of robot tracklets,
Tr_h: Set of human tracklets,
Tr'_h: Set of recent human tracklets,
K_{up}: Update threshold,
K_{acc}: Accuracy threshold
Ensure: $\psi_{sv}(\cdot, W_{sv})$: Social neural network
1: **if** $|T_t| \bmod L_{\text{trak}} = 0$ **then**
2:    $Tr_r \leftarrow$ Most recent robot tracklet ($L_{\text{trak}} \times s_r^s$)
3:    $Tr_h, Tr'_h \leftarrow$ Most recent human tracklets ($L_{\text{trak}} \times s_h^s$) $\times n$
4:    **if** $|Tr_r| \bmod K_{\text{up}} = 0$ **then**
5:       $Y_{Tr'_h} \leftarrow \text{label}(Tr'_h)$
6:       $\hat{Y}_{Tr'_h} \leftarrow \psi_{sv}(Tr'_h, W_{sv})$
7:       $Tr'_h \leftarrow \emptyset$
8:       **if** $\text{binary\_acc}(Y_{Tr'_h}, \hat{Y}_{Tr'_h}) < K_{\text{acc}}$ **then**
9:          $Y_{Tr_h} \leftarrow \text{label}(Tr_h)$
10:         $Y_{Tr_r} \leftarrow \text{label}(Tr_r)$
11:         $D_{\text{new}} \leftarrow [Tr_h, Y_{Tr_h}] \cup \{(x, y) \in Tr_r, Y_{Tr_r} \mid y = 0\}$
12:         $W_{sv} \leftarrow \text{train}(D_{\text{new}})$
13:         $Tr_r \leftarrow \emptyset$
14:         $Tr_h \leftarrow \emptyset$
15:       **end if**
16:    **end if**
17: **end if**

sets are cleared (lines 13 and 14) after each update, preventing the model from becoming overly specialized to previously observed data. The hyperparameters L_{trak}, K_{up}, and K_{acc} require careful tuning for optimal performance.

4.7 Conclusion

This chapter presented a comprehensive overview of Robot Online Learning (ROL), a paradigm that empowers robots to autonomously learn and adapt during operation, eliminating the need for direct human intervention. We began by introducing the core concept of ROL and subsequently articulated its significance within the broader field of robot learning. We argued that ROL represents a crucial step towards truly autonomous robots and offers a means to overcome inherent limitations of traditional offline learning methods. We then identified two key challenges that must be addressed to realize the full potential of ROL: first, the challenge of autonomous sample extraction, enabling robots to glean valuable learning information directly from their sensor data; and second, the critical issue of mitigating catastrophic forgetting during long-term online learning. To address these challenges, we presented several promising approaches. For autonomous sample extraction, we discussed methods based on positive–negative (P–N) learning and knowledge transfer. To combat catastrophic forgetting, we introduced a gate-controlled Long Short-Term Online Learning (LSTOL) architecture. Finally, we illustrated the practical application of ROL by demonstrating its effectiveness in enhancing socially-compliant robot navigation.

While the research presented in this chapter showcases encouraging progress in the field of autonomous robot learning, significant challenges remain. One particularly pressing issue is the development of a robust mechanism for robots to autonomously determine convergence during online learning and subsequently maintain stable model performance. A key obstacle in addressing this challenge stems from the fundamental premise of ROL: the absence of human intervention. This lack of explicit ground truth makes it difficult for the robot to evaluate the real-time performance of its learning model, to ascertain whether learning has converged, and to decide when further model updates are necessary to maintain stability. Currently, the robotics community lacks a definitive solution to this crucial problem. Existing research primarily focuses on theoretical analyses grounded in Lyapunov theory and offline evaluations of ROL models. Future work must directly confront this challenge, developing practical methods for autonomous convergence detection and stable model maintenance in long-term ROL scenarios. This will be a critical step in transitioning ROL from theoretical promise to real-world deployment, enabling robots to learn and adapt continuously and reliably in dynamic and unpredictable environments.

References

1. Yan, Z., Duckett, T., Bellotto, N.: Online learning for human classification in 3D LiDAR-based tracking. In: Proceedings of the 2017 IEEE/RSJ International Conference on Intelligent Robots and Systems (IROS), pp. 864–871. Vancouver, Canada (2017)
2. Yan, Z., Duckett, T., Bellotto, N.: Online learning for 3D lidar-based human detection: experimental analysis of point cloud clustering and classification methods. Auton. Robot. **44**(2), 147–164 (2020)
3. Yang, R., Yan, Z., Yang, T., Ruichek, Y.: Efficient online transfer learning for 3D object classification in autonomous driving. In: Proceedings of the 2021 IEEE International Conference on Intelligent Transportation Systems (ITSC), pp. 2950–2957. Indianapolis, USA (2021)
4. Yang, R., Yan, Z., Yang, T., Wang, Y., Ruichek, Y.: Efficient online transfer learning for road participants detection in autonomous driving. IEEE Sens. J. **23**(19), 23522–23535 (2023)
5. Shackleton, J., Voorst, B.V., Hesch, J.A.: Tracking people with a 360-degree lidar. In: Seventh IEEE International Conference on Advanced Video and Signal Based Surveillance (AVSS), pp. 420–426 (2010). https://doi.org/10.1109/AVSS.2010.52
6. Teichman, A., Thrun, S.: Tracking-based semi-supervised learning. Int. J. Robot. Res. **31**(7), 804–818 (2012). https://doi.org/10.1177/0278364912442751
7. Dewan, A., Caselitz, T., Tipaldi, G.D., Burgard, W.: Motion-based detection and tracking in 3d lidar scans. In: D. Kragic, A. Bicchi, A.D. Luca (eds.) IEEE International Conference on Robotics and Automation (ICRA), pp. 4508–4513 (2016). https://doi.org/10.1109/ICRA.2016.7487649
8. Okunevich, I., Lombard, A., Krajnik, T., Ruichek, Y., Yan, Z.: Online context learning for socially compliant navigation. IEEE Robot. Autom. Lett. **54**, 1–8 (2025)

References

9. Yan, Z., Sun, L., Duckett, T., Bellotto, N.: Multisensor online transfer learning for 3D lidar-based human detection with a mobile robot. In: Proceedings of the 2018 IEEE/RSJ International Conference on Intelligent Robots and Systems (IROS), pp. 7635–7640. Madrid, Spain (2018)
10. Yang, R., Yang, T., Yan, Z., Krajnik, T., Ruichek, Y.: Preventing catastrophic forgetting in continuous online learning for autonomous driving. In: Proceedings of the 2024 IEEE/RSJ International Conference on Intelligent Robots and Systems (IROS), pp. 1–8. Abu Dhabi, UAE (2024)
11. McCloskey, M., Cohen, N.J.: Catastrophic interference in connectionist networks: the sequential learning problem. Psychol. Learn. Motivat. **24**, 109–165 (1989)
12. Ratcliff, R.: Connectionist models of recognition memory: constraints imposed by learning and forgetting functions. Psychol. Rev. **97**(2), 285 (1990)
13. Rolnick, D., Ahuja, A., Schwarz, J., Lillicrap, T., Wayne, G.: Experience replay for continual learning. Adv. Neural Inform. Process. Syst. **32**, 1925 (2019)
14. Kirkpatrick, J., Pascanu, R., Rabinowitz, N., Veness, J., Desjardins, G., Rusu, A.A., Milan, K., Quan, J., Ramalho, T., Grabska-Barwinska, A., et al.: Overcoming catastrophic forgetting in neural networks. Proc. Natl. Acad. Sci. **114**(13), 3521–3526 (2017)
15. Schwarz, J., Czarnecki, W., Luketina, J., Grabska-Barwinska, A., Teh, Y.W., Pascanu, R., Hadsell, R.: Progress and compress: a scalable framework for continual learning. In: Proceedings of the 35th International Conference on Machine Learning (ICML), pp. 4535–4544 (2018)
16. Zenke, F., Poole, B., Ganguli, S.: Continual learning through synaptic intelligence. In: Proceedings of the 34th International Conference on Machine Learning (ICML), pp. 3987–3995 (2017)
17. Li, Z., Hoiem, D.: Learning without forgetting. IEEE Trans. Pattern Anal. Mach. Intell. **40**(12), 2935–2947 (2018). https://doi.org/10.1109/TPAMI.2017.2773081
18. Lopez-Paz, D., Ranzato, M.: Gradient episodic memory for continual learning. In: Annual Conference on Neural Information Processing Systems (NeurIPS), pp. 6467–6476 (2017)
19. Ramapuram, J., Gregorova, M., Kalousis, A.: Lifelong generative modeling. Neurocomputing **404**, 381–400 (2020). https://doi.org/10.1016/J.NEUCOM.2020.02.115
20. Wang, H., Xiong, W., Yu, M., Guo, X., Chang, S., Wang, W.Y.: Sentence embedding alignment for lifelong relation extraction. In: Conference of the North American Chapter of the Association for Computational Linguistics: Human Language Technologies (NAACL-HLT), pp. 796–806 (2019). https://doi.org/10.18653/V1/N19-1086
21. Sherstinsky, A.: Fundamentals of recurrent neural network (RNN) and long short-term memory (LSTM) network. Phys. D **404**, 132306 (2020)
22. Chen, C., Liu, Y., Kreiss, S., Alahi, A.: Crowd-robot interaction: crowd-aware robot navigation with attention-based deep reinforcement learning. In: IEEE International Conference on Robotics and Automation (ICRA), pp. 6015–6022 (2019). https://doi.org/10.1109/ICRA.2019.8794134
23. Chen, Y.F., Everett, M., Liu, M., How, J.P.: Socially aware motion planning with deep reinforcement learning. In: IEEE/RSJ International Conference on Intelligent Robots and Systems (IROS), pp. 1343–1350 (2017). https://doi.org/10.1109/IROS.2017.8202312
24. Yang, Y., Jiang, J., Zhang, J., Huang, J., Gao, M.: St2: Spatial-temporal state transformer for crowd-aware autonomous navigation. IEEE Robot. Autom. Lett. **8**(2), 912–919 (2023). https://doi.org/10.1109/LRA.2023.3234815
25. Kalal, Z., Matas, J., Mikolajczyk, K.: P-N learning: bootstrapping binary classifiers by structural constraints. In: IEEE Conference on Computer Vision and Pattern Recognition (CVPR), pp. 49–56 (2010). https://doi.org/10.1109/CVPR.2010.5540231
26. Kalal, Z., Mikolajczyk, K., Matas, J.: Tracking-learning-detection. IEEE Trans. Pattern Anal. Mach. Intell. **34**(7), 1409–1422 (2012). https://doi.org/10.1109/TPAMI.2011.239
27. Mohler, B.J., Thompson, W.B., Creem-Regehr, S.H., Pick, H.L., Warren, W.H.: Visual flow influences gait transition speed and preferred walking speed. Exp. Brain Res. **181**(2), 221–228 (2007)
28. Pan, S.J., Yang, Q.: A survey on transfer learning. IEEE Trans. Knowl. Data Eng. **22**(10), 1345–1359 (2010). https://doi.org/10.1109/TKDE.2009.191

29. Burgard, W., Moors, M., Fox, D., Simmons, R.G., Thrun, S.: Collaborative multi-robot exploration. In: IEEE International Conference on Robotics and Automation (ICRA), pp. 476–481 (2000). https://doi.org/10.1109/ROBOT.2000.844100
30. Thrun, S.: Learning metric-topological maps for indoor mobile robot navigation. Artif. Intell. **99**(1), 21–71 (1998). https://doi.org/10.1016/S0004-3702(97)00078-7
31. Moravec, H.P.: Sensor fusion in certainty grids for mobile robots. AI Mag. **9**(2), 61–74 (1988)
32. Read, J., Bifet, A., Pfahringer, B., Holmes, G.: Batch-incremental versus instance-incremental learning in dynamic and evolving data. In: Proceedings of the 11th International Symposium on Advances in Intelligent Data Analysis (IDA), pp. 313–323 (2012). https://doi.org/10.1007/978-3-642-34156-4_29
33. Saffari, A., Leistner, C., Santner, J., Godec, M., Bischof, H.: On-line random forests. In: Proceedings of the 12th IEEE International Conference on Computer Vision Workshops (ICCV Workshops), pp. 1393–1400 (2009). https://doi.org/10.1109/ICCVW.2009.5457447
34. Chen, Y.F., Liu, M., Everett, M., How, J.P.: Decentralized non-communicating multiagent collision avoidance with deep reinforcement learning. In: IEEE International Conference on Robotics and Automation (ICRA), pp. 285–292 (2017). https://doi.org/10.1109/ICRA.2017.7989037
35. Cosar, S., Donatiello, G., Bogorny, V., Gárate, C., Alvares, L.O., Brémond, F.: Toward abnormal trajectory and event detection in video surveillance. IEEE Trans. Circuits Syst. Video Technol. **27**(3), 683–695 (2017). https://doi.org/10.1109/TCSVT.2016.2589859
36. Cho, K., van Merrienboer, B., Gülçehre, Ç., Bahdanau, D., Bougares, F., Schwenk, H., Bengio, Y.: Learning phrase representations using RNN encoder-decoder for statistical machine translation. In: Conference on Empirical Methods in Natural Language Processing (EMNLP), pp. 1724–1734 (2014). https://doi.org/10.3115/V1/D14-1179
37. Adeli, V., Adeli, E., Reid, I., Niebles, J.C., Rezatofighi, H.: Socially and contextually aware human motion and pose forecasting. IEEE Robot. Autom. Lett. **5**(4), 6033–6040 (2020). https://doi.org/10.1109/LRA.2020.3010742
38. Kretzschmar, H., Spies, M., Sprunk, C., Burgard, W.: Socially compliant mobile robot navigation via inverse reinforcement learning. Int. J. Robot. Res. **35**(11), 1289–1307 (2016). https://doi.org/10.1177/0278364915619772
39. Vintr, T., Yan, Z., Eyisoy, K., Kubis, F., Blaha, J., Ulrich, J., Swaminathan, C.S., Mellado, S.M., Kucner, T., Magnusson, M., Cielniak, G., Faigl, J., Duckett, T., Lilienthal, A.J., Krajnik, T.: Natural criteria for comparison of pedestrian flow forecasting models. In: Proceedings of the 2020 IEEE/RSJ International Conference on Intelligent Robots and Systems (IROS), pp. 11197–11204. Las Vegas, USA (2020)
40. Okunevich, I., Hilaire, V., Galland, S., Lamotte, O., Shilova, L., Ruichek, Y., Yan, Z.: Human-centered benchmarking for socially-compliant robot navigation. In: Proceedings of the 2023 European Conference on Mobile Robots (ECMR), pp. 1–7. Coimbra, Portugal (2023)

Chapter 5
Conclusions and Perspectives

Abstract This chapter summarizes the full text and gives prospects for future research.

Keywords Mobile robotics · Robot perception · Robot learning · Human-aware navigation · Long-term autonomy

5.1 General Conclusions

Intelligent mobile robotics has witnessed remarkable advancements over the past two decades. This book has focused on two crucial aspects: robot perception and robot learning, with downstream applications in human-aware robot navigation and long-term robot autonomy. It began by addressing the critical, yet challenging, problem of benchmarking in mobile robotics (see Chap. 2). The discussion encompassed three key areas: standardizing evaluation procedures, developing common testbeds, and creating open datasets. It then presented our work on robot perception (see Chap. 3) and robot learning (see Chap. 4), respectively. It is important to note that these two research areas are both independent and complementary. Robot learning relies on the data provided by robot perception, while the knowledge acquired through learning can, in turn, enhance the robot's perceptual capabilities. Our focus in robot perception was on the use of 3D lidar and the processing of resulting point cloud data. The corresponding chapter provided a review of 3D lidar, considering its operating principles, scanning architectures, physical characteristics, data representation, and industrial applications. Subsequently, it introduced object detection and tracking techniques based on 3D lidar data, including a representative method, "adaptive clustering", which demonstrated a favorable balance between processing speed and clustering accuracy compared to similar methods at the time. In the realm of robot learning, this book explored the Robot Online Learning (ROL) paradigm, discussing its motivation, challenges, illustrative examples for object detection, a method for mitigating catastrophic forgetting, and an approach to improve socially-compliant robot navigation.

5.2 Research Perspectives

ROL is undoubtedly a research direction ripe for continued exploration. As noted in Chap. 4, ROL faces the inherent challenge of "lack of ground truth" due to the absence of human intervention. A promising avenue for addressing this challenge lies in the synergistic combination of Online Learning (OL) and Reinforcement Learning (RL), recognizing the robot as an active agent within its environment. This approach can be viewed as a natural extension of the research presented in Sect. 4.6, but with a tighter coupling between OL and RL. Specifically, this integrated framework would enable the robot to acquire high-confidence learning samples through interaction with the environment [2] and to evaluate the performance of learned models online using real-time feedback. It is important to acknowledge that this active learning paradigm necessitates careful consideration of robot ethics, including, but not limited to, safety, privacy, and accountability.

Another compelling research direction involves endowing robots with the capacity for intuitive prediction or "guessing". While human-human interaction can often be seamless, as exemplified by collaborative work in a logistics warehouse, the introduction of mobile robots can introduce complexities and ambiguities.[1] One potential contributing factor is the current lack of human-like intuition in robots, such as the ability to anticipate or guess, a crucial driver of instantaneous and smooth interaction. Guessing is a fascinating human cognitive activity, a creative process blending rationality and what might appear as irrationality. Crucially, guessing is not arbitrary, it is often informed by prior learning. If robots are to be granted this capability, connectionist-based approaches may prove particularly fruitful.

Finally, engineering considerations in robotics warrant continuous attention. While the robotics community may not have reached a consensus on whether engineering falls squarely within the realm of research, this book advocates for the view that the essence of robotics lies in system integration. This perspective is supported by researchers at Berkeley, who have highlighted the significant gap between AI and systems, emphasizing the rich potential for research in this area [1]. The emergence of *IEEE Robotics and Automation Practice* further underscores the community's growing recognition of this issue. Enabling model deployment on edge devices is of paramount importance. Beyond facilitating long-term robot autonomy, edge deployment can also circumvent numerous data privacy concerns. Furthermore, we can explore the integration of ROL with federated learning to further enhance privacy. Achieving these objectives requires explicit consideration of the limited computational resources and low power requirements of robots, posing significant challenges to algorithm optimization and software-hardware integration.

---

[1] https://yzrobot.github.io/navware/.

References

1. Stoica, I., Song, D., Popa, R.A., Patterson, D.A., Mahoney, M.W., Katz, R.H., Joseph, A.D., Jordan, M.I., Hellerstein, J.M., Gonzalez, J.E., Goldberg, K., Ghodsi, A., Culler, D.E., Abbeel, P.: A Berkeley view of systems challenges for AI. CoRR https://arxiv.org/abs/1712.05855 (2017)
2. Yan, Z., Sun, L., Krajnik, T., Duckett, T., Bellotto, N.: Towards long-term autonomy: a perspective from robot learning. In: Proceedings of the AAAI-23 Bridge Program on AI and Robotics, pp. 1–4. Washington, USA (2023)

Index

A
Adaptive clustering, 8, 48, 49, 51–54, 61, 73, 95
Artificial intelligence, 1
Autonomous driving, 11, 12, 25, 28, 30, 31, 33, 43, 45, 46, 48, 61, 71

B
Bounding boxes, 33, 34, 52–54

C
Catastrophic forgetting, 8, 69, 71, 80, 91, 92, 95
Concept drift, 79

D
Deep neural network, 34, 45, 48

E
Edge devices, 47
Embodied intelligence, 1, 2, 4, 7
Embodied learning, 67, 68
Embodied perception, 3, 25, 28, 39, 46
Euclidean distance, 49, 54, 56, 88, 90
Exteroceptive sensors, 44

G
Ground truth, 17, 19, 52–54, 83, 92, 96

H
Hand-crafted features, 8, 48, 54, 55, 61
High-definition maps, 45
Human-aware robot navigation, 4, 20, 22, 95
Human-robot interaction, 20, 86

K
Knowledge transfer, 8, 72, 77, 79, 86, 92

L
Lifelong learning, 68
Long-term deployment, 69, 86
Long-term robot autonomy, 3, 4, 39, 44, 95, 96

M
Mahalanobis distance, 59
Mobile robotics, 7, 23, 33, 35, 40, 44, 47, 48, 55, 58, 61, 68–70, 95
Multi-modal perception, 45
Multi-robot systems, 2, 13, 17

O
Object detection, 6, 14, 17, 28, 30, 34, 44, 46–48, 52, 58, 61, 69–71, 80, 86, 95
Object tracking, 33, 58
Open science, 5
Operational design domain, 12, 68

P
Point cloud clustering, 28, 34
Publish-subscribe model, 25

R
Robot ethics, 96
Robot online learning, 4, 55, 68, 91, 95

S
Sensor fusion, 28
Social acceptability, 20, 34, 89
Software engineering, 4, 5, 7, 22
Streaming data, 58
System integration, 96

MIX
Papier aus verantwortungsvollen Quellen
Paper from responsible sources
FSC® C105338

If you have any concerns about our products,
you can contact us on
ProductSafety@springernature.com

In case Publisher is established outside the EU,
the EU authorized representative is:
**Springer Nature Customer Service Center GmbH
Europaplatz 3, 69115 Heidelberg, Germany**

Printed by Libri Plureos GmbH
in Hamburg, Germany